TAP IN

Liberate Your Time, Save Money, and Transform Your Life with AI

TAP IN

Liberate Your Time, Save Money, and Transform Your Life with AI

A Simple Guide to Using Artificial Intelligence Every Day for Everyone

PAUL JOFFE

Printed in the United States of America

Published by Igniting Souls
PO Box 43, Powell, OH 43065
IgnitingSouls.com

LCCN: 2024919011
Paperback ISBN: 978-1-63680-377-7
Hardcover ISBN: 978-1-63680-378-4
e-book ISBN: 978-1-63680-379-1

Available in paperback, hardcover, and e-book.

Access free resources, including many more tips on AI, applicable uses, and the latest in AI news

Become part of our community and share your curiosities, questions, and concerns.
We are all in this together.
JoffeHouse.com

TABLE OF CONTENTS

SECTION IV: PRODUCTIVITY

SECTION V: ANYONE CAN CREATE

SECTION VI: FUN AND FAMILY

SECTION VII: ACCESSIBILITY

FOREWORD

I grew up with Paul, cousins who lived a few miles away from each other during the dawn of the home computing age. But neither of us imagined ever going deep into tech. I became a TV news reporter and documentary filmmaker. Paul worked in film, interactive learning, and games. The internet boomed and busted, and then the next generation arose: Web 2.0, powered by social platforms and the smartphone.

As technology became more and more pervasive in our lives, Paul emerged as a leader in how it intersected with entertainment, education, and business. He built innovative products and companies that reached tens of millions of people. I never anticipated my world and Paul's would converge, but with the AI revolution upon us, each of us was separately leaning into it and developing a passion for sharing what we've learned.

I launched the social media brand Demystify AI, posting short videos to amplify how AI is increasingly relevant in our daily lives. Then, I learned that Paul was pursuing a similar mission. He called me to share that he had written a new book with the goal of opening access to this most incredible technology of our time to as many people as possible. He asked me to review it before publication, and I was blown away by the level of detail and valuable advice it contains to encourage us all to embrace AI to improve our efficiency and effectiveness.

With *Tap In: Liberate Your Time, Save Money, and Transform Your Life with AI*, Paul has put together a highly accessible, impactful, and enlightening guide that enables any of us, regardless of technical expertise, income, or education, to unlock life-changing opportunities with Artificial Intelligence. The book is a gateway to understanding how AI can be a powerful ally in solving everyday problems, offering cost-effective and time-saving solutions that can enhance our quality of life. Whether you're looking to reach your fitness goals, handle household repairs, manage your finances, or boost your productivity, every chapter provides simple step-by-step directions and the right tools to deploy. None take more than a few minutes to get started.

I personally chose to focus on demystifying AI because of the wide gap between the corporations and experts that are rapidly deploying it, and most of us who are struggling to grasp its implications and what it means. Paul doesn't just talk about AI in abstract terms or big ideas; he demonstrates its practical use through real-world examples. You'll learn how AI can be your Personal Coach, Nutritionist, and Assistant Chef. He reveals how AI can make you an Author, an Entrepreneur, or simply plan your next vacation with little stress and significant cost savings. Why should saving money, or making money, be only for big tech and financiers? AI is here for all of us.

The book is a testament to Paul's deep knowledge, empathy, and unwavering commitment to helping others harness the power of technology. He masterfully breaks down complex AI concepts into easy-to-understand language. Whether you are a novice or a seasoned tech enthusiast, you'll find insights and actionable strategies to enhance your life.

We are living in an unprecedented era where the wonders of AI are transforming our daily lives in ways we once only imagined in science fiction. I encourage you to dive into this book with an open mind and a willingness to explore the possibilities. You will find yourself energized and empowered to tap into AI's full potential and the invaluable opportunity to transform your life.

Happy reading, and welcome to the future!

Lisa Bernard
Founder/Host @lisabernard.demystifyai on Instagram

INTRODUCTION

I work in tech and have for over twenty-five years. When Artificial Intelligence broke into the mainstream with the release of Open AI's ChatGPT-3, it was not a shock but an inevitability. Yet, as my colleagues and I played with the new wonder toy, even we were struck by how easy and accessible the power of AI had become. We were astonished at how the straightforward chatbot interface had turned AI into something anyone could access. *Finally*, we thought, the democratization of technology was truly here. Everyone would have the power of thousands of computers and millions of human minds from today and history.

Yet a funny thing happened as AI chatbots and a slew of new AI tools have exponentially advanced productivity, creativity, finance, and health (to name a few). Machine Learning engineers have benefitted. Businesses have embraced it. Marketers are all over it. Financial markets have fallen in love with it. But a large part of the population approaches it like a poisonous snake, a beast of intimidation, fear, and distrust.

I use the creature metaphor intentionally. People do not view AI as a machine, tool, or object. They see it as something alive, unpredictable, and practically autonomous. That creates a barrier to getting close to it, familiarizing yourself with it, and understanding it. It also makes it alien—something people cannot and would not know what to do with.

Time and again, at dinners with friends and families, the topic of AI comes up, and people frown, squirm, or roll their eyes. Many of these are highly educated people whose work (as almost all workplaces are today) is flooded with technology. While some remain quiet, even disdainful, others are eager to express their concerns and doubts. They're trained by the headlines highlighting massive impending job loss, copyright infringement galore, creative thievery at every turn, energy-sucking machines, and (imminently) runaway autonomous drones and robots. If the technology itself is sci-fi, how large swaths of people are reacting to it is living up to science fiction's dystopic dreams.

For some period, I gamely joined in the debate around the looming doom so many have come to embrace. I am unabashedly optimistic about AI and happily indulge in arguments around it. I am not shy about evangelizing the incredible benefits it can bring to our society, such as curing diseases, finding solutions to clean energy, democratizing education, and reducing costs on everything from the food supply to transportation to home prices. AI is a technology of scale that can reach many. Someday, it will help us unlock a world of abundance.

But what about today, and how about things a little more tangible and closer to home? Soon after ChatGPT-3 launched, I began playing with AI tools beyond how I used them in my workplace. I was so amazed at how accessible it had become that I wanted to see its magic applied to whatever need I had: meal prep, health diagnosis, home fixes, and, of course, questions—questions, questions, and more questions. The superintelligence in the pocket was always my ever-ready informant.

I knew it would start showing up in more and more places, such as fitness, nutrition, and travel, where a digital economy fierce with competition was bound for disruption. I jotted notes to keep track of my experimentation and became conscious of the time and money AI was saving me. I changed my approach to conversations about AI. Instead of arguing how it would transform medicine, I offered practicalities—AI could prepare them for a doctor's visit or improve their credit score.

Then, a little something happened. Unlike AI as a future world changer, people were interested in and open to hearing about it. The poisonous snake was turning into a pet turtle—still a bit exotic but far more friendly and approachable.

The more time I spent researching AI tools and leveraging AI's benefits, the more excited I became about what this could mean for everyone. I wanted to talk about it more, tell people about it, and make sure everyone knew the ridiculous returns they could get from it. It didn't take much effort and wasn't at a low cost, or even at no cost—*it would save them money*—a lot of it.

I want everyone to know, so I've written this book. Like a friendly chatbot, I've aimed to make it easy to access, quick to digest, highly useful, and offer a little something for everyone. I'm quite confident there is a lot for everyone.

You'll encounter concrete examples, practical applications, and a blueprint for integrating AI into your daily life. This book is not just a guide; it's an invitation to view the world differently, with AI as an empowering companion rather than a threatening replacement. You will find specific uses, tested tools, and accurate calculations on how AI can improve your life. Much of this comes in a form we are all already familiar with: mobile and web apps with quick and free access. However, this is a fresh breed of app, directly tied into the groundbreaking artificial intelligence engines reshaping the world and ushering in a new age.

By the end, you won't just understand the transformative power of AI; you'll know how to harness it to improve your life. If you are reading this, the extraordinary journey to learn how to navigate this bold new world for your benefit starts here. This is not a path to replace human ability but to enhance it, giving you more freedom of time, finance, creativity, and capability. It's about improving the quality of our lives, allowing us more space to grow, learn, and engage with the world around us.

Don't worry; you won't lose your identity or soul to the machines. But you can discover how to get much more out of life and yourself. AI can save you time and money, boost your emotional well-being, improve your relationships, and open new opportunities. Let's find out how right now.

SECTION I

HOME

CHAPTER 1

BECOME A DIY SUPERSTAR

One might not immediately think of AI as a trusted toolbox companion, yet it is quietly revolutionizing the Do-It-Yourself category. We all know the frustration of household appliances suddenly belting out an odd sound, followed by long hours spent deciphering repair manuals or searching the web for the exact issue you're confronting.

Now, picture a world where AI identifies issues through sound recognition or visual analysis, offers step-by-step repair instructions, and even estimates the cost and time saved by doing it yourself. Not only can AI demystify repairs, but it can also empower us to tackle challenges head-on with newfound confidence.

A Plumber Whenever You Need It

It's 3 a.m., and a hiss awakens you. It's not a snake lurking in your bathroom but a misbehaving toilet mechanism. For many, a plumbing issue like this would mean hours spent searching the web, reading through DIY forums, or calling an expensive plumber. Some will ignore it as long as possible while frustration and water bills rise. It might even cause damage to the bathroom.

The challenge is identifying the root cause and figuring out a solution without breaking the bank or waiting days for a professional.

If you have used chatbots in the past, such as for customer support, they have likely been constrained and incompetent at understanding your queries. That has all dramatically changed since the release of ChatGPT and other new-generation chatbots, such as Claude and Genesis. These are often called LLMs, which stand for Large Language Models.

LLMs are incredibly good at understanding natural language descriptions and, more impressively, precisely answering questions with clarity and detail. You can even show them pictures or ask them to make one!

Whether text, image, or audio, what you input into a blank chatbot field acts as a "prompt." You will hear the word "prompt" regularly when it relates to AI tools, and this means you are asking a question or providing material for the AI to evaluate.

Let's see how the chatbot can quickly solve the problem of the hissing toilet—no matter the day or time! We'll use ChatGPT by Open AI for this one, an LLM you've almost certainly heard of.

Step-by-Step: Solving the Case of the Hissing Toilet

AI Tool: ChatGPT

1. Request Help

Open the chatbot, and in the empty text box, type in your question just as if you were speaking it:

My toilet is making a hissing sound; what is the problem?

2. Review Responses

In my case, the response was a list of five potential issues with short explanations and fixes. Different prompts can provide varied responses, but the natural language capability of today's AI makes responses very close to how you'd expect a person to respond, even with different wording and tone.

3. Drill in and Get the Answer

Rather than researching each one, I decided to let it do the work:

What is the most likely cause of the hiss?

It replied immediately:

The most likely cause of a hissing sound in a toilet is a faulty fill valve. This valve controls the flow of water into the toilet tank, and over time, it can wear out, become misaligned, or get clogged, leading to the hissing noise as water continuously flows or leaks through it. Adjusting or replacing the fill valve is often the solution to this issue.

I prompted:

How do I replace a faulty fill valve? Show me a video or diagram.

ChatGPT responded:

I found a video that provides a detailed guide on how to replace a faulty toilet fill valve. You can watch it for a step-by-step visual demonstration. Here's the link:

How to Replace a Toilet Fill Valve | Toilet Repair | The Home Depot.

After a quick review of the short and concise video, I had one final ask. I took a photo of my toilet bowl, inserted it into the prompt field, and asked ChatGPT what type of fill valve I needed. It replied:

The image shows a modern toilet tank with a fill valve on the left side. When looking for a replacement, you should look for a "universal toilet fill valve" designed to fit most toilets. These are adjustable and can be set to the correct height for your toilet tank. Brands like Fluidmaster and Korky offer popular options widely available at home improvement stores.

Most of you will have a local hardware store or supply store such as Home Depot or Lowe's in mind. But you could provide your address and then request the location of the nearest store. It will instantly find the address and phone number of a store in your neighborhood, and you're on your way to a quick, cheap fix.

DIY Beyond a Noisy Toilet

One excellent outcome of interacting with AI chatbots for home repairs is the knowledge and self-empowerment that's gained. I used to procrastinate undertaking a home repair project. I dreaded all the hours spent searching manuals or the Internet for the right solution. Inevitably, the instructions on how to solve the issue were not quite suitable for my situation or model.

Now, I'm excited to take on the task. Not only does handling things yourself feel great, but it also has a long-term impact on the belief and confidence in yourself to solve problems that were once daunting and eventually costly.

Here is how AI can help you with home repairs and improvement.

1. **Instant Guidance and Troubleshooting**
   - Chatbots provide step-by-step guidance for a wide range of home repair tasks. Whether fixing a leaky faucet or patching up a hole in the wall, they can guide you through the process.
2. **Highly Personalized Advice**
   - Chatbots tailor advice based on the specific details of your project, such as the materials you're working with, the tools you have available, or the particular models you're fixing. The prompts can include photos, drawings, and sound recordings.
   - Often, a chatbot will inquire about information to provide the most accurate and productive response.

3. **Safety Tips**
    - Safety is paramount in DIY projects. Chatbots provide tips and best practices to prevent accidents or injuries.
    - They can also tell you if certain chemicals or other materials have special handling requirements and what to do if exposed.
4. **Material and Tool Recommendations**
    - Every job has preferred materials and tools for your specific repair, which will get you the best results. Ask your chatbot what those are to save time and money!
    - Don't have the exact right thing? A chatbot can advise on alternative tools or materials if you're missing something or want to purchase something less expensive.
5. **Locate a Store Near You**
    - Provide your address, and the chatbot will tell you the closest place to find your items.
6. **Access to Tutorial Videos and Articles**
    - Chatbots will direct you to resources for a more in-depth explanation or visual guide.
7. **Cost Estimation**
    - They can provide the estimated costs for any required tools and materials, itemized part by part.
8. **Project Management**
    - For larger tasks, a chatbot can organize your project by breaking it down into manageable

steps, reminding you of timelines, and helping you track your progress.

9. **Community Support**
    - Want to connect with others in the DIY community or find forums to participate in? A chatbot can point you to places that are invaluable for getting additional advice, sharing your experiences, or finding inspiration for future projects.

Many of these benefits, as with other elements in this book, are available through search on the World Wide Web. However, the time AI chatbots save you and the precision of their answers and information—free from advertising and sponsored results—is truly amazing. It will make you far more willing and able to take on DIY projects and empower you in numerous other areas that can transform your daily life.

Savings

The emotional and educational benefits of DIY are great, but at the end of the day, we are all highly conscious of our bank accounts. In many instances, AI's most important and tangible impact is on our pocketbooks. Problems like the noisy toilet can be solved with the free version of a chatbot. That's right—all this knowledge and assistance is absolutely free.

Additionally, unlike the current internet search, which provides far less usable and actionable information, there are no sponsorships or paid advertising to distract you from the task at hand.

Let's look at how AI did in saving money and other impactful elements such as time.

Money

- **Cost of a Plumber Visit:** On average, plumbers charge between $50 and $200 for a service call. And that's to identify the issue. Depending on the problem, repairs might add another $100 to $500.
    - **Savings**: *$200* for the visit, parts, and repair.
- **Water Wastage:** A deteriorated fill valve leak can waste up to 6,000 gallons of water monthly. There is a significant cost to doing nothing.
    - **Savings:** Water costs an average of $0.005 per gallon, leading to a loss of *$30 per month* or *$360 per year.*

Estimated Net Cost Saved: $180—the plumber's visit less the cost of the fill valve.

If you are a procrastinator like many of us and let the toilet leak water, the savings could be *hundreds of dollars a year more.*

Time

- **Research and Dabbling**: The diagnosis and solution took a few minutes, saving at least an hour of searching manuals or the Internet and fiddling around with your toilet bowl.
    - While this is a definite benefit, especially in the middle of the night, in future examples, we'll see

how AI tools can reduce a multitude of hours, days, and even weeks to minutes.

Headaches

Yes, avoiding headaches is a definite saving! We'll see time and time again how AI, with its speed, accessibility, and preciseness, can completely alter your state of mind. It will make you not only wealthier but healthier.

Empowerment Starts Here

The hissing toilet incident and the demonstration of how artificial intelligence can support DIY begins our journey into how AI will massively improve your daily life. This improvement is available to everyone's benefit, wherever you are and with whatever means. It generally requires nothing more than the simple internet-connected device that we already keep with us 100 percent of the time—our phone.

Let's keep going!

CHAPTER 2

THE FOOD WHISPERER

Let's move from the bathroom to what should be a much more alluring place to enjoy the benefits of AI—the kitchen. Unfortunately, for many of us, mealtime can be as intimidating and exhausting as a home repair. Even for those who love cooking, deciding what to make with whichever ingredients are available on a given day while also meeting the dietary requirements of yourself or your household drains away the pleasure of preparing meals.

AI as a culinary assistant is massively powerful. In just a couple of minutes, it can offer a recipe with step-by-step cooking directions based on the ingredients you have. It can recommend the recipe based on cuisine preference, dietary restrictions, and desired cooking time. Gluten-free, vegetarian, keto, vegan, low-carb, high protein—no matter your regime, AI will rapidly offer delicious and nutritious options for exactly the items you have.

Even better—no cooking experience is required.

The Nightly Meal Trial

We've all been there many times. On a regular workday in March, we arrive home after dark and open the fridge to a random assortment of ingredients staring back at us.

Ordering takeout again would be a snap, but it costs a minimum of twenty to thirty dollars on one meal and involves a thirty-to-forty-minute wait, pushing us later and later into the evening. On some nights, winging it with ingredients in the kitchen can be a bit of fun, but it tends to miss far more often than hit.

There used to be limited options in moments like these. Sure, the internet is filled with recipes, but finding one that fits your current situation of time and ingredients, never mind taste and diet, takes a lot of effort. In my case, I'm usually halfway through before I realize an ingredient is missing.

For many of us, the power of the LLM chatbots is enough to solve the cooking dilemma. It's wild how quickly you can have multiple options at your fingertips. Like with repairs, all this in-depth knowledge and capability is available with the free versions.

Step-by-Step: Solving the Nightly Meal Trial

AI Tool: ChatGPT

1. Activate AI

I popped open ChatGPT on my smartphone and tapped the microphone.

2. Review Potential Ingredients

Over the next minute, I went through what was in my fridge and cabinets, prompting the chatbot with my natural voice just like this:

I would like to cook a meal. Here are the ingredients I have. Onions. Carrots. Celery. Apple. Mozzarella cheese. Parmesan cheese. Cheddar cheese. Eggs. Chicken breast. Lettuce. Soy sauce. Ketchup. Mayonnaise. Italian dressing. Brazilian nuts. Almonds. Sour cream. Yogurt. Tomato sauce. Salt. Pepper. Bagels. Pita bread. Frozen peas. Frozen beans. Frozen broccoli. Canned coconut milk. Canned garbanzo beans. Canned black beans. Raisins. Spaghetti. Rice. Crushed canned tomatoes. Canned green chilies, diced. What can I make in less than thirty minutes for dinner?

I know it's a completely random assortment of ingredients that would drive most of us to the delivery app, and maybe we'll even despair when we realize how much more we'll spend on a mediocre meal *again* that night.

Yet, within a few seconds after my prompt, magic happened.

For a quick and tasty dinner, how about a chicken stir-fry with a side of garlic bread? the chatbot replied.

Okay, I thought, not bad. But I'd had a lot of meat that week.

Can you give me something vegetarian? I responded.

Tick tick, a few more seconds, and back came:

How about making a savory vegetable frittata? You could also make a side salad with lettuce, chopped apple,

and almonds, with Italian dressing. Does this idea sound good to you?

Inspired, I pushed further. I asked for something gluten-free and low in sodium, and it offered **a Hearty Vegetable and Bean Soup with a side of Yogurt-dressed Salad**.

Then I proposed vegan, and it suggested ***a Spicy Black Bean and Vegetable Rice Bowl.***

3. **Request Recipe**

   I asked it to provide me with the three recipes and instructions, and within a few seconds, they were all listed.

There it was. In less than five minutes, I had three recipes using just the available ingredients and fitting various diet considerations. With one short phrase, "Give me other options," I could have had countless more recommendations meeting any conditions I appended.

Want a meal in five minutes instead of thirty? It recommended a ***Vegetable and Cheese Pita***.

Beyond the Quick Solve

The example is a bit dreary, but AI can be more than a quick fix in a tight spot. It can provide massive inspiration and opportunities in the kitchen. Let's look at the ways you can leverage this capability:

1. **Explore Different Cuisines from Around the World**
   - Craving something exotic? Discover dishes from different cultures, complete with recipes,

historical context, and even wine pairings. It's like having a culinary global tour guide in your kitchen.

2. **Learn How to Cook Anything**
   - Get step-by-step instructions, video tutorials, and even real-time guidance. Picture yourself making a delicious béchamel sauce, a saffron aioli, truffle risotto, or Coquilles St. Jacques. AI acts as a virtual chef, guiding you step by step through a recipe. Made a misstep? Just tell it what happened, and it will provide a solution.
3. **Transform the Taste of Your Favorite Things**
   - Feeling adventurous? Ask for a recommendation on adding twists to a favorite recipe or combining preferred foods. Pair chocolate with beetroot for a cake, add sumac to a chicken marinade, or matcha to your brownies.
4. **Make Shopping Easy**
   - AI will compile a shopping list based on your meal plans, ensuring you buy only what you need. Some apps can even sync with local grocery stores to check item availability, compare prices, or arrange for home delivery.
5. **Sustainability and Reduction in Food Waste**
   - AI promotes a lower carbon footprint by suggesting seasonal and locally available ingredients. It can also offer tips on conserving energy in the kitchen, from efficient cooking methods to appliance usage.

Improve Health and Wellness

This is such a large and important category that I've separated it from the list. AI can be incredibly beneficial in this area. Here's one simple example: you're making a smoothie. Simply prompt the AI about what could make it healthier.

In my case, it suggested adding flax seeds for Omega-3 or swapping in almond milk for lower calories. These minor, intelligent tweaks to a healthier diet can make a huge difference in your health day-to-day. Such health benefits are immeasurable.

If you asked an AI chatbot for recipes, snacks, and adjustments to meals to reduce sodium in your diet, it would provide endless options with these benefits:

- Lower Blood Pressure
- Decreased Risk of Heart Disease
- Reduced Risk of Stroke
- Prevention of Osteoporosis
- Weight Management

Once again, this is all available for free! No paging through large cookbooks, reading long articles, or guessing about what will help you the most. Your food well-being companion is with you all the time.

Savings

Let's look at our savings on that March evening when AI quickly facilitated a healthy meal.

Money

- **Meal Cost**: Getting takeout averages $25 after all costs, such as tips and fees, versus $8 for the ingredients at home.
    - **Savings**: *$17*

Net Money Saved: *$170 per month* if done only ten days a month, and *hundreds more* if you make food at home daily.

Time

- **Time Difference**: Takeout requires a forty-minute wait on average.
    - **Savings**: five to thirty minutes, depending on what you make.

Net Time Saved: *ten to thirty-five minutes.*

Additional Savings

Here are a few more additional savings just that one small moment provided:

- Less waste on delivery packaging materials
- Use of existing ingredients that often go bad
- A healthier meal that also supported my weight loss plan
- Stress reduction

Tools and Tips

Sometimes ChatGPT is the only sidekick you need in the kitchen, but there are various apps customized for shopping, recipe logging, and cooking. For those looking for a robust AI-powered app to help manage the entire cycle of planning, shopping, and cooking regularly, there are some good options below.

- **ChatGPT** helps you decide what to cook based on the ingredients you have on hand.[1]
- **Be My Chef** recipe generator offers step-by-step cooking directions tailored to your dietary preferences and ingredient availability.[2]
- **Yummly** provides personalized recipe recommendations based on your tastes and dietary restrictions, with the added benefit of creating shopping lists and meal plans.[3]
- **Paprika Recipe Manager** manages recipes, creates meal plans, and generates grocery lists. It allows you to scale recipes and keep track of pantry items.[4]

AI, Cooking, and Life Hacking

I could go on indefinitely about how AI can transform your meals, diet, and shopping. We may not all become Michelin chefs, but we can gain the knowledge and skills

1 https://chat.openai.com/

2 https://letsfoodie.com/ai-recipe-generator/?ref=futuretools.io

3 https://www.yummly.com/

4 https://www.paprikaapp.com/

to make food gathering and consumption—the most basic human needs—a productive, healthful, and joyful experience every day.

As we complete the second chapter on our journey of how AI empowers our daily lives, it's worth highlighting one of the most valuable benefits of interacting with an AI chatbot or app: it remembers you! The more you prompt it, whether for recipes, nutrition, diet, or simply inspiration, the better it learns your preferences and evolves to become your custom Super Assistant.

CHAPTER 3

CULTIVATE A GREEN THUMB

Gardening is a cherished hobby for many and a source of sustenance for others. It connects us with nature and provides a sense of accomplishment when we see our plants thrive. However, it can also be daunting due to the unpredictable elements, pests, and diseases that affect our plants. This is where AI transforms gardening into a more manageable and rewarding endeavor. AI can help novices and seasoned gardeners make informed decisions, save time, and reduce waste.

Veggie Tale

Consider my neighbor Sarah, an enthusiastic but struggling gardener. She noticed her tomato plants were developing yellowing leaves with dark spots, a sign of distress. Not knowing whether it was due to overwatering, a nutrient deficiency, or a disease, she felt overwhelmed. The prospect of losing her crop after investing so much time and care was disheartening.

Step-by-Step: Saving the Tomatoes

AI Tool: Picture This App

1. Getting AI Help

Sarah decided to leverage AI technology to diagnose her plants' condition. She downloaded the *Picture This* app, a highly recommended tool for gardeners. *Picture This* and other apps like it use AI to identify plants and diagnose diseases, pests, and nutrient deficiencies through photo recognition. Downloading and signing up for the app takes less than a minute.

2. Diagnosis and Recommended Treatment

Within moments of taking a picture of her tomato plant, the app analyzed the photo and identified the problem as early blight, a common fungal infection in tomatoes. It provided a detailed description of the disease, its symptoms, and its impact on plants.

It further suggested several organic treatment options, emphasizing the importance of acting quickly to prevent the spread. One recommended solution was applying a homemade fungicide mixture consisting of baking soda, liquid soap, and water.

3. Implementing the Solution

Sarah followed the app's organic treatment recommendation to minimize chemical use in her garden. She prepared the fungicide mixture following the exact directions from the app:

- One tablespoon of baking soda
- One-half teaspoon of liquid soap
- One gallon of water

She mixed these ingredients in a spray bottle and applied them to the affected leaves, ensuring they were coated thoroughly on the top and underside as directed.

4. Ongoing Monitoring

The app assessed the garden's moisture levels to prevent future outbreaks and suggested pruning the plants to improve air circulation around them.

While a moisture meter could be helpful, Sarah found the app incredibly effective at evaluating the plant's watering levels. It suggested increasing water with an exact amount and schedule and even allowed her to set up phone notifications to remind her!

5. Recovery and Maintenance

Over the following weeks, Sarah monitored her tomato plants closely, applying the baking soda solution weekly, following the reminders about watering, and using the *Picture This* app to assess their health regularly.

The Outcome

Thanks to the AI-driven diagnosis and the application of recommended treatments, Sarah's tomato plants recovered from the early blight. Her proactive measures, guided by AI insights, turned her struggling garden into a thriving one. Much to her delight, she harvested a healthy crop of tomatoes later that season.

All the Amazing Things AI Can Do in Gardening

1. **Plant Recognition**
   - Snap a photo of a plant, and AI can identify the species and provide care instructions, making garden maintenance easier.
2. **Disease and Pest Identification**
   - AI analyzes photos of your plants to identify diseases and pest infestations, offering immediate solutions.
3. **Care Instructions**
   - Based on the plant, your location, sunlight information, and analysis of leaves and branches, it provides detailed care guidance and support with time-sensitive notifications.
4. **Soil Analysis**
   - If you're more serious and have soil sensors, AI can analyze the data and recommend soil amendments to optimize your garden's health.
5. **Watering Optimization**
   - Use AI to anticipate your garden's watering needs based on weather forecasts and soil moisture levels, helping to conserve water.

6. **Yield Prediction**
    - AI can predict crop yields for those growing fruits and vegetables, helping plan for storage or sale.
7. **Garden Planning**
    - AI algorithms can suggest plant placements based on sunlight, space, and companion planting guidelines to maximize garden health and yield.

Savings

Sarah's use of AI in gardening as she battled early blight in her tomato plants illustrates significant savings in both time and money. Let's break down these savings to understand the benefits of integrating AI into gardening practices.

Time

- **Disease Identification**: Researching plant symptoms online or consulting a gardening expert could take several hours to days, especially for an inexperienced gardener. Sarah identified the disease in minutes.
    - **Savings**: Approximately *two to three hours* (or more) of research time.

- **Solution Research**: After identifying the problem, finding and deciding on a treatment method could take additional hours of research. AI immediately provided a solution.
    - **Savings**: At least another *one to two hours* of research and decision-making time.

- **Ongoing Monitoring and Adjustment**: Without precise guidance, continuous monitoring and adjustments might be more frequent and time-consuming. AI advice efficiently targeted Sarah's efforts.
    - **Savings**: Likely *two hours* over the growing season.

Estimated Net Time Saved: *six to eight hours* for this single incident, *multiplied over a year* if managing a garden or collection of house plants where additional incidents are likely.

Money

For specialized tasks like gardening, AI-powered apps built for the specific field will be much more powerful. While most apps have a free tier for some basic identification, you'll want to subscribe for the full features. Still, as you'll see below, this will save you money!

- **Cost of Incorrect Treatments**: Without an accurate diagnosis, Sarah might have spent money on general pesticides or fungicides that were not suited to her specific problem.
    - **Savings**: $20 to $50 on ineffective treatments.
- **Plant Replacement:** Losing tomato plants to early blight would require purchasing new plants or seeds, not to mention the lost time and resources already invested in the affected plants.
    - **Savings**: $2 to $5 per plant, totaling $20 to $100, depending on the size of her garden.

- **Water and Resource Conservation:** By optimizing watering through AI guidance, Sarah could save on her water bill and the cost of excess fertilizers.
    - **Savings**: Even a ten percent reduction in water and fertilizer use can save $10 to $30 over a growing season in a small to medium garden.

Estimated Net Money Saved: *$50, which assumes an app subscription of* $35 to $50 per year. You are also protected against the unpredictable costs of future new issues. The app costs don't increase regardless of whether you have one problem or twenty.

Of course, all your plants would benefit from the power of AI insights and solutions for nothing more! You'll increase your confidence and horticultural prowess, provide help to neighbors and friends, and gain the opportunity to learn and grow more.

Environmental and Health Savings

While more challenging to quantify, AI-guided organic treatment solutions like Sarah's baking soda mixture help reduce the environmental impact of chemical fungicides and pesticides. This approach supports a healthier garden ecosystem and reduces exposure to harmful chemicals for the gardener and local wildlife.

Tools and Tips

Many apps are designed to identify, diagnose, and support plant care. Here are some especially simple and effective options. Take a picture, and the apps provide identification, detailed information, an instant assessment of health, and easy-to-follow tips to help with care.

The first two are best used with their subscription tier. The third is free, although it is designed for crops, so its capabilities are more limited for house plants.

- **Picture This**[5]
- **Plantum**[6]
- **Plantix** is a free Android-only app specific to crops. It would work well for a vegetable garden like my friend Sarah's. Plantix supports farmers all over the world in caring for their crops.[7]
- Numerous free videos about plant care are available, and like with most search-based efforts, AI chatbots like ChatGPT and Gemini will find them for you in seconds with a simple prompt, such as "How Do I Care for my Tomato Plant?"

A Partnership Between Nature and AI

Sarah's experience highlights the power of AI in transforming gardening practices. By integrating AI tools, gardeners can quickly diagnose and treat common issues, saving their plants and ensuring bountiful harvests. It not only saves time and resources but also promotes sustainable gardening practices.

Sarah's story is also a testament to the combination of people and AI. AI can't garden itself, but it can teach people how to make gardening more enjoyable, productive, and sustainable. This partnership between technology and humans can improve every one of our futures.

---

5 https://www.picturethisai.com/

6 https://myplantum.com/

7 https://plantix.net/en/

SECTION II

HEALTH AND FITNESS

CHAPTER 4

MEDICAL CLARITY

The internet has been an incredible boon for instant access to health information. It contains a trove of data about conditions and treatments and provides us with large amounts of detail in our online medical records. Nothing could be more personalized and empowering than having such valuable information. Yet it's also incredibly daunting. What does it all mean? Who do we trust? Where do we even begin?

AI tools are designed to demystify medical jargon, interpret test results, and suggest items to raise with your doctor. This knowledge leads to a deeper understanding of one's health, fostering a partnership approach to medical care rather than a passive patient experience.

Medicine is the perfect example of a field that benefits immensely from AI but cannot replace the role of expert practitioners. Engaging actively with AI-driven insights allows patients to approach medical appointments confidently, equipped with the right questions and a better understanding of their health indicators. It enhances your patient-doctor relationship, providing a path to clear, actionable insights that lead to better health outcomes.

Your Test is Abnormal, Now What?

A close friend of mine recently found himself unusually fatigued with terrible mood swings. His doctor ordered some blood tests, and what came back via a friendly email notification from his health system was a bundle of results that were abnormal and annotated with loads of incomprehensible terms. He had an upcoming appointment that was still a week away and was desperate to understand what he was confronting.

He came to me seeking some help. The tests indicated an "abnormal thyroid function test," focusing on irregular TSH, T4, and T3 results. It's pretty much impossible to know what things mean without a medical degree, so I suggested we turn to AI.

Step-by-Step: Understanding Thyroid Function Tests

AI Tool: Claude

1. First Objective

We wanted to understand the results. Most free tiers of LLMs can provide this information with high accuracy. For this request, we used Claude, an LLM with a free tier for information inquiry.

What is a thyroid function test, and what do abnormal results indicate? my friend prompted.

In clear and straightforward language, Claude instantly provided the following:

- A clear explanation of the thyroid function test
- Normal and abnormal ranges of the critical indicators TSH, T3, and T4
- What higher or lower results in the ranges could indicate
- Suggested next steps

2. Digging In

In less than sixty seconds, we received extremely helpful information! Based on this data, my friend's test results indicated **hypothyroidism**.

For a few more minutes, we queried AI a bit further:

What are the symptoms of hypothyroidism?

What causes it?

What are the treatments?

What lifestyle changes can help?

What questions should I ask my doctor?

Claude returned these simple questions with equally clear but informative answers. What used to take hours of reading through different websites as your head spun was now quickly consumable and comprehensible.

3. Trusting the Results

Unlike using the web, the results were not shrouded in uncertainty about whether the information had been "promoted" to our eyeballs by a self-interested third party wanting our business. In an internet search, a company pays to get to the top of the list, regardless of whether it has the best or most truthful information. There are no sponsored results with AI models; they lay out the pure scientific facts plainly.

LLMs have been trained on *all* the data that exists on the internet. This goes far beyond what websites have chosen to publish or the information companies pay for you to see. The most valuable part of LLM training in medicine is the vast library of medical books and articles that doctors study in their training. Did you know AI has been able to pass the written portion of medical exams, such as the United States Medical Licensing Examination?

4. Understanding Breeds Calm and Confidence

Lab results and medical terms may initially seem hard to understand, but at the end of the day, they're usually numbers with normal and abnormal ranges. They're very easy to comprehend, with a little lift from an AI chatbot!

5. Attend the Appointment

On the day of my friend's doctor's appointment, he was armed with information and questions. He felt in control and less anxious about the visit. He was a participant in his healthcare rather than a lab rat. Knowledge is incredibly empowering, providing greater command over essential decisions in your life. Such insights are now highly accessible to all of us and available exactly when we need them.

Ten Ways to Own Your Medical Care

AI has a transformative impact when it comes to supporting one of the most critical parts of your life. It can quickly and succinctly give you a roadmap for managing any medical condition you face. Among other things, this includes:

1. Symptom evaluation
2. Demystifying test results
3. Simplifying medical jargon
4. Providing questions and topics to discuss with your doctor
5. Recommending where to find specialists
6. Suggesting treatment options
7. Delivering insight into medications and their effects on you
8. Making recommendations on lifestyle changes and diet to improve your condition
9. Connecting you to support groups or forums for people with similar experiences
10. Understanding long-term implications to your health and what should be monitored

Savings

Money

- **Healthcare Bills**: Unnecessary follow-up appointments, tests, or incorrect treatments can be costly.
    - **Savings:** *Hundreds to thousands of dollars* in medical expenses are avoided through a proactive approach that can lead to earlier and more precise interventions.

- **Reduced Medication:** Patients can avoid unnecessary or less effective medications by accurately understanding health conditions and treatments.
    - **Savings:** Varies depending on insurance but can save *hundreds of dollars.*

Net Money Saved: *Hundreds of dollars* in appointments, tests, and medications.

Time

- **Research:** Usually involves hours, spread over several days or weeks, as you wait for appointments and follow-ups.
    - **Savings:** Condensed into *minutes.*

- **Doctor's Visit Preparation:** Ease of preparation and specificity of questions lead to more efficient and focused doctor visits.
    - **Savings:** *Hours* of planning, traveling, and attending follow-up appointments.

Net Time Saved: *About ten hours.*

Additional Savings

- **Reduced Anxiety and Emotional Cost**: The immediate access to information helps reduce the anxiety that often accompanies unclear medical test results. Being informed and more in control of your situation reduces the emotional toll of health issues.
- **Improved Health Outcomes**: With a better understanding of a condition, you can make informed decisions about your health, leading to quicker and more effective management of a condition. This can improve overall health outcomes, which is invaluable.

Tools and Tips

- **Symptom Checkers**: Tools like Ada Health and Healthily offer user-friendly interfaces for symptom input and provide reliable analysis.[8]
- **Test Result Analysis**: AI platforms like ChatGPT and Claude AI can interpret medical test results and suggest next steps.[9]
- **Doctor Finder Services**: Websites like HealthTap and ZocDoc help you find specialists suited to your specific symptoms and potential conditions.[10]

---

8 ada.com; http://www.livehealthily.com

9 https://claude.ai/chats

10 http://www.healthtap.com; http://www.zocdoc.com

The Power of Knowledge

Medical care is a perfect example of how combining AI and people can have an incredible impact. Both patients and doctors are empowered with the rapid and in-depth analysis of symptoms, tests, and other data points that can lead to a comprehensive understanding of managing a condition most effectively. AI can continue to assist in tracking and comparing the results of future tests and providing clear and concise options each step of the way.

CHAPTER 5

ELEVATE YOUR FITNESS

Like medicine, fitness has benefited from the wealth of data collected and repackaged by modern technology to provide clear and concise action plans. It further shares medicine's biggest challenge: personalization. Whether it's for your body, diet, or schedule, we're bombarded with highly generalized plans and advice for standardization and mass consumption. Yet we know each of us is different. We need custom care for our bodies, lifestyles, and mental approaches.

This is where AI excels. There is no need for one-size-fits-all training and diet programs. AI provides personalized and continuously adaptive planning to help you achieve your goals.

A Plan Perfect for You

AI-powered apps can help you make a diet and training plan in minutes, keep you on the plan with reminders, and adjust based on your feedback and data. The apps can be supplemented with watches such as a Fitbit™ or Apple Watch®, but those can be expensive and aren't required to unlock the magic of AI.

For most fitness apps, the initial steps after signing up only take a few minutes and produce a workout plan for your particular goals, lifestyle, and schedule. Here's what a typical flow looks like:

Step-by-Step: Curating Your Fitness Journey

AI Tool: Fitness App of Choice

1. Enter Personal Info

Elements such as age, weight, and height determine your body mass index (BMI).

2. Define Your Objective

Being specific helps the app tailor routines for your individual focus and generally includes identifying the following:

- Weight Loss
- Strength
- Muscle Tone
- Fat Loss
- Sport Fitness
- And more...

3. Specify Goals

Setting targets, such as targeting a specific weight, BMI, or strength index, helps refine the regimen.

4. Timing

How quickly do you want to achieve your goals? Maybe you have an upcoming wedding, holiday gathering, or vacation at the beach.

- Set your target for six weeks, three months, or whatever works for you, and the AI tool will tailor your daily fitness plan to meet the date.

5. Input Your Experience

Your fitness training background helps further refine your plans, such as years, daily or weekly amount, and overall level assessment (e.g., beginner to advanced).

- The workout will be catered to your needs, whether you are a neophyte or a fitness maven.
- All information is kept private, although some apps encourage sharing your progress to get support from friends and the community. This is entirely optional.

6. Location and Equipment Access

Knowing where workouts occur and with what apparatus is another highly customizable part of your fitness blueprint.

- Identify whether you want to integrate gym equipment, free weights, a treadmill, a stationary bike, or other machines into your training.
- You can say "no" to all equipment, and the app will generate a plan you can use anytime, anywhere—just you and Earth's gravity.

7. Workload

How many days do you want to work out?

- Depending on other goals, this will determine the length and intensity of the workout.
- If you have specific goals within a set amount of time, the app will suggest what is required (or recommend that you change the timing of your goal).

8. Set Notifications

You will be reminded of the day and time you want to work out, keeping you in a steady routine.

As you can see, these fitness routines become incredibly personalized to your needs and schedule. Given so many variables, a trainer would have difficulty making the perfect plan for you, but AI does it in seconds.

Over time, the app will adapt the training regimen instantly based on data such as your success in keeping the routine, weight changes, etc.

Syncing your Apple Health or Google Fit data, all free with your smartphone operating system, provides loads of helpful data to the apps, assuring a greater hyper personalization never previously available until AI arrived for all of us.

An Expansive Role in Your Health and Conditioning

Let's see all the ways AI is revolutionizing fitness:

1. **Personalized Workout Plans**
    - AI analyzes your fitness level, goals, and preferences to create a customized workout plan. It can adjust your strategy based on progress, ensuring you're constantly challenged without being overwhelmed.
2. **Nutritional Guidance**
    - In the next chapter, we'll see how AI provides personalized nutritional advice to support your fitness goals.
3. **Real-Time Form Correction**
    - Through motion tracking and analysis, AI can offer real-time feedback on your exercise form, reducing the risk of injury and improving the effectiveness of your workouts.
4. **Personal Trainers**
    - AI-driven virtual trainers—or smart matches made with human trainers—deliver motivation support, real-time guidance during workouts, and instant feedback.
5. **Recovery Insights**
    - By monitoring your activity and rest periods, AI suggests optimal recovery times and practices, preventing overtraining and enhancing performance.

6. **Mental Wellness**
    - A slew of AI-powered apps can provide mindfulness and mental wellness exercises tailored to your stress levels and psychological state, recognizing the importance of mental health in achieving physical fitness goals.
7. **Supplement and Hydration Recommendations**
    - Based on your activity level, diet, and goals, AI can recommend when and what supplements to take and how much water to drink.
8. **Sleep Analysis**
    - AI monitors your sleep patterns and provides insights and recommendations for improving sleep quality, acknowledging its critical role in fitness and recovery.
9. **Social Motivation**
    - Get recommendations to fitness platforms that connect you with like-minded individuals for challenges and support, offering inspiration and fostering a sense of community. If you need local, real-world suggestions, those, too, are quickly available.

Each of these functionalities highlights AI's capability to provide a comprehensive, integrated approach to fitness, encompassing physical exercise, nutrition, mental health, and recovery.

Savings

Money

- **Personal Trainer**: The average cost of a personal trainer ranges from $40 to $70 per hour, depending on location and the trainer's expertise. However, users can save significantly by using AI-powered fitness apps, which generally have subscription costs of around $10 to $30 per month.
    - **Savings**: *$400 to $800 per month* if you had three sessions per week with a personal trainer, which would cost $480 to $840.

- **Gym Memberships:** Traditional gym memberships can range from $20 to over $100 monthly. AI fitness programs include home workouts that can eliminate the need for a gym membership.
    - **Savings:** *$50* based on the average gym membership cost.

- **Equipment and Travel Costs:** AI fitness solutions offer minimal or no equipment options, eliminating costs associated with purchasing gym equipment. Additionally, by working out at home, users save on travel expenses to and from a gym or training facility.
    - **Savings:** *Hundreds of dollars per year.*

Net Estimated Money Saved: Even with the cost of an app subscription, the savings are, *at minimum, hundreds of dollars* and, for many people, *thousands per year.*

Time

- **Travel Time:** By eliminating the need to go to a fitness center, users can save an average of thirty minutes to one hour per workout session.
    - **Savings:** *One and a half to three hours per week* for three weekly sessions.

- **Scheduling Flexibility:** AI fitness apps allow for workouts anytime, so users do not have to align their schedules with gym hours or trainer availability.
    - **Savings:** This flexibility can save significant time that might otherwise be wasted adjusting daily schedules and allows you to maintain workouts when you travel.

- **Speed to Results:** AI-powered fitness solutions offer personalized workout plans that adapt to users' progress and fitness levels.
    - **Savings:** Faster and more targeted results than generic training programs.

Overall, using AI in fitness provides significant cost and time savings and offers a highly personalized, flexible, and engaging approach to maintaining physical health.

Tools and Tips

- There are numerous fitness apps for any device. While they almost universally have subscription requirements, most come with free trials. Try a few to find the right one, and remember to cancel the subscription before the free trial ends. You can quickly do this in your app settings.

- Some specific apps you might try and support most of the functionality stated in this chapter include:
    - **Fitbod**
    - **Shred**
    - **Workout by Fitness 22**
    - **Evolve AI**[11]

- New apps constantly emerge, adapting to the latest AI technology advancements. Use the free tier of any web-connected AI chatbot, such as **Gemini** or **Bing Co-Pilot**, to search for the most popular and well-rated apps at any given time.
- You can also make specific requests in the type of app you want, such as virtual or real trainers, focus on weight loss (or other goals), and cost.

A Leap Forward in Personal Health Technology

With excellent bespoke planning accessible to everyone, AI has forever changed our ability to stay fit. By following a structured, AI-powered plan, you can achieve goals more efficiently and sustainably. These intelligent tools adapt to your progress, ensuring your journey is as dynamic and unique as you are. Whether you're just starting or looking to enhance your fitness routine, AI offers the insights and support needed to navigate your path to wellness.

---

11 https://www.fitbod.me; https://www.shred.com; https://www.fitness22.com; https://www.evolveai.com

CHAPTER 6

NUTRITION AND HEALTHY LIVING

Nutrition is fundamental to our health, impacting everything from daily energy to long-term well-being. Incorporating AI into our dietary practices allows for personalized nutrition advice that adapts to individual health needs, lifestyles, and fitness goals.

This chapter explores how AI is transforming our approach to diet and nutrition, making it easier to maintain a healthy lifestyle and meet our health and wellness objectives.

Tailored Diet for Any Lifestyle or Goal

Want to lose ten pounds for an upcoming family event? Maybe a wedding, anniversary, or major birthday celebration?

How about getting in better shape for the summer and maintaining throughout and even beyond? Maybe now is the time to shed some weight and keep it off.

Want to be certain to keep to your specific dietary needs? Or try a new regimen with a different focus or balance of foods?

By leveraging AI tools, we can adhere to personalized, efficient dietary plans that include calorie counting, food types, and eating patterns without the time-consuming aspects of traditional dieting. This fits seamlessly into a busy lifestyle and ensures we remain consistently engaged and motivated throughout the weight loss journey.

The Latest Tools Have Leveled Up

Like with fitness, a mobile app is the best way to leverage AI for nutrition. These apps track multiple aspects of your daily eating habits, especially focusing on calorie input and weight management.

These apps require more time investment by users than many other categories in this book, as they depend on daily inputs of what you eat for each meal. However, the insights are extensive, and if you are serious about regulating what you eat, managing or losing weight, and keeping a healthy diet, they are precious. Most apps also recommend what to eat at any meal and provide recipes.

There are numerous app options, and many have been around for years—just like diets! AI has been integrated more extensively in recent years, including analyzing and applying learnings from all the data the apps have captured over the years.

One of the most significant areas of AI-powered advancement is image recognition accuracy. This allows you to take pictures of your meal and have the app analyze it for the type and quantity of food in your meal, making the input and tracking process significantly easier.

Step-by-Step: Setting Up for a Healthier Life

AI Tool: Nutrition Mobile App of Choice

1. App Selection

Like with fitness apps, you should test out a few options. Some apps require subscriptions, but most have a free trial. Be sure to look at reviews. Some of the more popular and well-featured ones are in the **Tools and Tips** section.

Most apps follow a similar approach, which I have outlined below so you can decide if this is for you. They do require meal-by-meal tracking of what you are eating. It's now supported by a significant advancement in AI image recognition, where snapping a photo of a plate of food and getting the lowdown on what you're about to eat is easy, fun, and accurate.

2. Personal Profile Creation

Apps need details on your measurements, age, gender, and objectives to personalize properly.

3. Weight Goal

Whether you want to lose, maintain, or gain weight is fundamental to the calculations that craft your plan. You can certainly have goals related to various nutrition-centered questions—sugar, sodium, carbs, and anything else you want to manage. However, the foundation of the analysis and plans is calorie counting, and calorie consumption is determined by target weight.

4. Timing

Most people have a target event or date range to achieve their goals, and the apps cater to these needs. Your requirements shape everything from calories consumed daily to the composition of the food you eat. The apps can help you manage it day by day with counters and graphs that give feedback on each meal and activity.

5. Nutrition

Healthy dieting is critically important, and with the help of AI analyzing years of data collected by sites and apps, this has become increasingly precise. AI apps can track the amount of protein, carbohydrates, fats, and other nutrients to the milligram.

With all these critical inputs added during a few minutes of onboarding, AI can now provide detailed meal plans, grocery lists, reminders, instant feedback, and adjustments.

A Day in the Life of an AI-Illuminated Diet

1. **Reminders** pop up on my phone at mealtime, keeping me on a regular eating pattern, which is crucial for metabolism management.
2. **Recommendations** are available if I haven't decided what to eat.

3. **Photo logging** is the quickest way to track what I eat. Let AI log the calories and nutrients automatically. If the numbers are off, it takes just a minute to tweak them.
    - Sometimes, I take pictures of my food at mealtimes and do the logging and tweaking at the end of the day.
4. **Feedback** immediately shows how I am tracking against the daily calorie and nutrient budget.
5. **My activity** is tracked by syncing with my smartphone fitness app, and my calorie allowance is adjusted in real-time based on how much I burn off.
6. **Progress bars** provide visual feedback on how I'm tracking toward weight loss goals, including graphs of weight change, calorie intake, and exercise.
7. **Personal rating** on meal satisfaction allows AI to tweak future meal suggestions, improving my contentment while maintaining caloric goals.
8. **Anticipate special occasions** like friends' nights out, and my overall day and week are adjusted in advance so I'm not thrown way off and playing catch-up towards my goals after a night of a little excess.
9. **Virtual Coaching** through push notifications delivers motivational tips and mini-challenges to keep me engaged.

Capabilities of AI in Nutrition

AI tools can extend diet and nutrition beyond calorie counting and meal tracking. Additional valuable functionalities include:

1. **Allergy Detection**
    - AI apps like **Spoon Guru** can analyze product labels and recipes to alert users about potential allergens.
2. **Supplement Recommendations**
    - Platforms like **Rootine** provide personalized vitamin and nutrient supplements based on DNA analysis, blood levels, and lifestyle data.
3. **Fitness Goal Alignment**
    - Apps such as those recommended in the previous chapter use AI to align dietary intake with workout regimes, optimizing nutrition based on exercise.
4. **Sleep and Energy Optimization**
    - Some AI-driven platforms suggest meal timings and compositions that enhance sleep quality or boost energy levels throughout the day.

Savings

Money

- **Daily Meal Planning:** Apps often suggest recipes based on sales or promotions at local grocery stores. Opting for seasonal and sale items can save approximately $2 to $5 on groceries daily.
    - **Savings:** *$1,825 annually.*

- **Reduced Food Wastage:** By buying only what you need for planned meals, you can reduce food wastage, which typically saves about $5 to $10 per week.
  - o **Savings:** *$520 per year.*
- **Nutritional Consultations:** The average cost of visiting a professional nutritionist or dietitian can range from $70 to $100 per session. The costs can be significant, assuming a monthly consultation is needed for ongoing dietary management.
  - o **Savings**: *$420 to $600 per year.*
- **Weight Loss Programs:** Commercial weight loss programs can cost anywhere from $20 to $50 per month for basic online plans to more than $100 for comprehensive plans, including meals.
  - o **Savings**: *$240 to $1,200 per year.*
- **Experimentation with Supplements:** Unnecessary supplementation can be costly, with individuals often spending $20 to $100 monthly. In some cases, supplements may be unnecessary; in others, money is poorly spent on the wrong ones.
  - o **Savings**: *$240 or more annually* with AI evidence-based supplement advice tailored to individual needs that avoids unnecessary products.
- **Health Risk-Related Cost Avoidance:** Poor Nutrition and Weight-Related Health Issues are the elephant in the room, costing money and potentially years of healthy living. Conditions like diabetes, heart disease, and joint problems due to obesity can lead to significant medical costs ranging from hundreds to

thousands of dollars annually in medications, treatments, and doctor visits.

Maintaining a healthy diet and weight can prevent these conditions, saving substantial amounts in healthcare. While difficult to estimate precisely, these savings can easily range from *hundreds to several thousand dollars annually*, depending on the severity and number of health issues prevented.

Estimated Net Annual Savings

- $1,825 from cost-effective shopping
- $520 from reduced waste
- $420 to $600 from nutritionist consultations
- $240 to $1,200 unnecessary supplements
- $240 to $1,200 from weight loss programs
- $500 to $5,000 or more from health risks

Not all apply to everyone, but it likely adds up to around *$3000 for most individuals* and more for families. This easily offsets the cost of app subscriptions, which run a reasonable $50-$100 annually.

Time

- **Meal Planning Efficiency:** Using AI-driven apps for meal planning can save approximately thirty minutes to one hour per day that might otherwise be spent searching recipes, calculating nutritional values, and planning meals.
    - **Savings:** *182.5 to 365 hours annually.*

- **Grocery Shopping:** Apps that generate shopping lists based on meal plans can save about twenty minutes per shopping trip by organizing lists by store layout or offering online shopping options.
    - **Savings**: *About thirty-five hours saved per year,* assuming two weekly shopping trips.

Estimated Net Time Saved: Two hundred-plus *hours per year.*

Tools and Tips

Here are some AI tools and resources that can assist with nutrition planning:

- **Foodvisor** stands out for its speed and accuracy in photo recognition. It makes keeping a food diary and assessing calories per meal super easy. Voice recognition is another helpful AI-enabled feature.[12]
- **MyFitnessPal** has an integrated fitness recommendation and functionality (as does Foodvisor), making dieting and working out easy to track in one place (and assuring your calorie counting allowance is balanced with your calories burned).[13]
- **Lifesum**, **Fooducate**, and **Lose It** are popular and robust calorie tracking and meal recommendation apps. Remember that the plans only work if you document your food consumption. The apps have

[12] https://www.foodvisor.io

[13] https://www.myfitnesspal.com

large searchable databases and barcode scanning for purchased groceries. Still, you need to estimate the size of portions to get the best accuracy in calorie counting.[14]

Get More From Your Food Pics

AI integrated with a nutrition app offers a highly personalized approach to diet management, allowing users to meet their health and fitness goals efficiently. Individuals can enjoy the benefits of a diet tailored specifically to their needs, contributing significantly to their overall health and reducing both time and financial costs.

This empowerment through technology and data-driven platforms improves dietary habits, enables healthy decision-making, builds self-esteem, saves significant money, and can significantly impact one's long-term health outcomes.

---

14 https://www.lifesum.com; http://www.fooducate.com; https://www.loseit.com/

SECTION III

PERSONAL FINANCE

CHAPTER 7

MANAGING YOUR MONEY

Navigating the complexities of personal finance, including budgeting, savings, credit, and expenses, can be daunting for many of us. AI is here to help! Like in other areas, it offers highly personalized insights, with the ability to automate savings, optimize and cut costs, help improve your credit, and detect fraudulent activities.

There was a time when you might need multiple apps to cover the plethora of financial improvements and optimizations available. The integration of AI has now allowed for all of this to be far more centralized, providing greater accuracy, convenience, and effectiveness. Let's see how it works.

Find Financial Freedom

If an app automates cost identification, it is AI-powered. Traditionally, finance apps have left a lot of work up to you. You would have to manually input specific items or identify what expense category they fall into, estimate your monthly or annual costs, answer endless questions, and organize the data on top of everything.

AI has streamlined the process, and if you have a genuinely AI-powered app, you should need no more than a few questions about your current financial situation and goals. By linking your bank and credit card accounts, hours of effort can be condensed into a fraction of the time. This indicates your platform will provide insights that can transform your financial management. So, get the right app!

Step-by-Step: The Foundation to Financial Freedom

1. Choosing Your App

There are recommendations in the **Tools and Tips** section. Still, the landscape is constantly changing, so I encourage you to use AI chatbots to search for the latest recommendations, as in other categories in the books. Regardless of which one you begin with, you should invest the time in trying a few apps with free trials to find one that is best for you. Key considerations are:

A. Choose an app with ten thousand or more positive reviews and a minimum rating of four and a half stars.

B. Make sure they allow a free trial period to test their functionality. It should cover all critical financial management requirements, ideally including:

- Budgeting
- Spend Tracking
- Lower Bills
- Savings Plan
- Credit Improvement
- Subscription Cancellation (you'd be surprised what you don't know you're paying for!)

C. Linking financial accounts is critical for AI to save you the time and money you deserve.

- The app should use a platform like Plaid, Mastercard, or another service that connects accounts to gather data only and not make any changes to a financial account.

D. If upon linking accounting you are confronted with manually adding, identifying, and sorting expenses (and not simply confirming), the platform is not using AI optimally.

Your investment in a Financial Management app will be more involved than other categories in this book, but it can also be the most impactful. You can save substantial money and time and put yourself on a path to financial stability and freedom.

2. Connect Accounts

Once you have chosen the app and completed the initial setup, connect all your financial institutions so AI can get the complete picture of your income, expenses, and savings (including investments). Connections should include:

A. Checking and Savings Accounts
B. Credit Cards
C. Investment Accounts, including 401k and Pensions
D. Mortgage and Loan Accounts

With this information, an excellent AI-empowered app will not only be able to automate identifying your income and expenses but also evaluate historical patterns, flag anomalies or potentially undesired expenses, and anticipate high and low periods of cash over the course of the year.

To ensure the best results, you will need to go through and confirm that items have all been properly categorized. This only must be done once, after which the app should be able to sort activity as it happens. AI will learn the names and types of places you receive income or incur expenses, educate you on your profile, improve categorizing, and understand spending patterns.

3. Budget and Savings

With everything set up, you can choose which objective to dig into. You will likely be queried during onboarding whether you are saving for retirement, planning to buy a home, etc. While this helps AI learn about you and customize it to your needs, the platforms can also handle multiple goals.

Let's look at an example. My friend Alex is a freelance graphic designer and struggles with budgeting and saving. With a fluctuating income, he finds it hard to keep track of expenses and save for future goals. The challenges this imposes on his life are not only financial. It adds tremendous stress to the relationship with his wife and family, leaving him in a persistent crisis as to whether he needs to change careers and lose the freedom of being a freelancer, which he cherishes.

Alex turned to an AI-powered finance app that evaluated his history of earning income, spending, and bills across each day of each month over previous years. It provided a monthly budget and detailed tracking so he could see how he was doing weekly. In months where he wanted to make adjustments, such as adding extraordinary expenses for a birthday or special occasion, his budget months before and after were immediately amended.

Most importantly for Alex as a freelancer, it recognized periods when he tended to have less work and income and others that were more consistent. This allowed him to save more in advance of potential dry periods.

4. Plan for a Large Purchase

One of the more challenging tasks for any person is saving up over the long term for a large purchase, such as a house. It can take years to accumulate the necessary down payment, and you don't want to cease spending on anything else during that period!

Smart finance apps can allow you to target multiple spending goals, determining the optimal amount to set aside for long-term purchases while allowing enough cash for nearer-term needs. They calculate the amounts you can set aside weekly or monthly, depending on your target purchase date. Apps generally don't have any control over your bank accounts, but some will provide an option to set up a savings account and withdraw a regular amount to ensure you are correctly saving. Others will recommend helpful tools such as monthly automated transfers into a savings account in your bank.

The Full Spectrum in Personal Finance

AI's role in personal finance extends far beyond budgeting and saving, encompassing a large swath of insights, controls, forecasting, recommendations, and notifications. These include:

1. **Tracking Income and Expenses**
   - AI will begin by monitoring your income and spending. Once it has this data, AI can help you analyze it and identify areas for improvement.
2. **Personalized Budgeting**
   - AI Algorithms can create customized budgets based on your spending history and financial goals. These budgets are more realistic and achievable than static budgets, and they can adapt over time as your financial situation changes.
3. **Goal Setting and Tracking**
   - AI helps set realistic savings goals and track your progress over time. This can help you stay motivated and on track.

4. **Forecasting Cash Flow**
   - AI analyzes income and spending patterns to predict your future cash flow. This can help plan for upcoming expenses and avoid overspending.
5. **Real-time Insights**
   - AI provides up-to-the-moment dashboards into your financial situation, helping you make informed decisions about spending and saving.
6. **Predictive Algorithms**
   - AI analyzes spending habits, estimates future expenses, and helps you plan accordingly.
7. **Personalized Guidance**
   - AI acts as a financial coach, providing tailored advice based on your financial goals and habits.
8. **Automated Savings**
   - AI determines how much you're able to save and set aside regularly.
9. **Debt Management**
   - AI suggests the best strategies to pay down debts, considering interest rates and repayment amounts.
10. **Fraud Detection**
    - AI monitors transactions for unusual activity, protecting individuals from fraud and identity theft.
11. **Financial Planning**
    - AI-powered virtual advisors provide personalized recommendations for retirement planning, insurance, and tax strategies.

12. **Credit Score Improvement**
    - AI analyzes credit reports and suggests actions to improve credit scores over time.
13. **Investment Guidance**
    - Through market data and trends, AI offers insights and investment recommendations.

Each of these tools leverages AI to make financial planning and saving easier and more effective, allowing you to achieve financial goals with greater precision and less effort. We'll visit a couple more of these in the coming chapters.

Savings

Money

- **Variability of Income**: The amount of dollar savings for better financial management differs significantly for each person's economic situation and goals.
    - For example, one who targets saving ten percent of their income at $50 thousand per year will be very different than if one's income is $150 thousand per year.
- **Investment versus Purchase**: The savings you invest will look very different from those you use for a purchase. For example:
    - If I am to save $20,000 to help cover new car payments over three years, then my savings are successfully converted into my needs.
    - If, instead, $20,000 over three years were saved to invest in retirement, then that $20,000 of savings

invested in an index fund with a conservative five percent compound interest over thirty years would be worth over $86,000. The person with this goal also met their objective thanks to AI, but the return is much greater.

- The actual average growth of the S&P 500 over the last thirty years has been ten percent. So, if you invested in an S&P Index fund and saw this type of return, you'd have almost $350,000 in thirty years! It makes you want to invest your car money and take the bus!

Net Money Savings: With the support of AI, it's easy to *save hundreds, if not thousands, of dollars annually* through financial management alone. This more than offsets the five to ten-dollar monthly subscription fee you'll pay.

If we add up the savings from every other chapter so far, it represents thousands of dollars altogether. We have plenty more to discover, and already the money savings are compiling, and your life is transforming!

Time

Time is another critical area of savings that continues to advance with AI-powered financial management. There is one thing in life in which we are all indisputable equals, no matter where we live, what we do, or how much we make: We all have only twenty-four hours in a day and seven days in a week.

Gaining more time each day to do what we choose, whether spending it with family, leisure, or productivity, can be a massively positive change in our lives. Let's save more!

- **Bill Payment and Transactions:** Automatically importing transactions saves approximately *five hours per month* compared to manual entry.
- **Budgeting and Review**: By streamlining budget planning and monthly reviews, you can save about *three hours per month* that you would otherwise spend on manual calculations and adjustments.
- **Financial Analysis and Future Planning:** Assuming we aim to gain savings and the opportunities it offers for vital purchases and investment returns, the speed at which AI can assist with the analysis and requirements to be successful is significant. A reasonable estimate is *four hours per month.*

Estimated Net Time Saved: *twelve hours per month, or sixty hours per year.* That's well over one week of standard working time, which is now free time for you to spend elsewhere.

Additional Benefits

- **Improved Financial Literacy:** AI tools often provide educational resources and personalized advice, enhancing your financial knowledge. This can lead to better financial decisions that compound savings.
- **Error Reduction:** AI reduces human error in financial management, such as missed payments or calculation errors, which can save money by avoiding late fees or financial missteps.
- **Enhanced Security:** AI can monitor accounts for unusual activity much more efficiently than humans, reducing the impact of financial fraud or

identity theft. Credit card companies use AI to detect anomalies, and it's time you got that power for yourself.

Tools and Tips

Financial management apps and software have existed for three decades, with major advancements over the last ten years, especially with mobile apps that can easily and securely connect to bank accounts and use features such as the camera to ingest financial data rapidly.

The more recent advancements in AI have only just begun to be incorporated. Many popular apps need to be optimized and involve considerable manual input and decision-making. As in other categories, check out multiple options and use AI chatbots or other tools to find the latest and greatest. Here are a few options to jumpstart your research:

- **Cleo** is the most AI-ready app we tested. It allows for advanced expense identification when accounts are linked and has an integrated chatbot that can guide you through and answer questions at any time.[15]
- **Albert** is another strongly AI-powered app that can quickly categorize expenses and offer budget suggestions. Its Smart Savings feature allows you to automatically deduct amounts regularly to keep for your goals without opening a new bank account.[16]

---

15 https://www.meetcleo.com

16 https://www.alberthq.com

- **You Need A Budget (YNAB)** and **NerdWallet** are comprehensive and popular apps that can support you in a wide range of financial planning. They have not integrated AI to speed onboarding and automated personal finance (such as expense identification) as others have, but I expect them to catch up in time.[17]

Look Out! Many of these apps have a base financial management subscription and additional subscriptions for more services. Others use upsells for credit cards or other financial instruments to offset their cost of app services.

Cash advance offers are a significant category that many apps focus on. These are based on the data the apps learn about you, your income history, and your spending habits. I have tried to avoid recommending these apps as they can cause you to get into debt. However, apps are updated regularly and change their offerings and business models, so please take care when accepting cash advances, credit cards, or other offers. They can increase your debt, harming your financial stability and putting your credit score at risk.

The next chapter is dedicated to credit scores, as credit rating is a crucial component of your financial health and worth investing in.

Taking Control of Your Money

AI in personal finance is a present-day reality that offers practical solutions for managing money more effectively. By automating mundane tasks, providing personalized advice, and enhancing decision-making, AI empowers us to achieve

---

[17] https://www.youneedabudget.com; https://www.nerdwallet.com

financial goals with greater ease and confidence. We can realize financial savings, fulfill purchase objectives, gain control over our financial lives, and build toward a future of economic stability and freedom. Personal finance management no longer needs to be intimidating or burdensome but rather a rewarding journey with excellent outcomes.

CHAPTER 8

IMPROVE YOUR CREDIT

A crucial aspect of personal finance is maintaining a healthy credit score, which is essential for securing loans, buying a home, and sometimes even employment opportunities. It enables freedom in life decisions such as starting a business or getting capital for other career-changing opportunities. A higher credit score can lead to lower interest rates and better terms on loans and will save you thousands, if not tens of thousands, of dollars over time.

However, navigating the intricacies of credit reports and scores can be daunting. Here, AI becomes a vital ally. By analyzing your credit report, AI identifies factors that pull your score down and suggests concrete steps to improve it. It can monitor your credit and alert you to potential fraud that will drag the rating down. Resolving credit errors can be a frustratingly long and belabored process, and having a bot keeping an eye on things, especially one that never sleeps or even blinks, is immensely valuable.

Step-by-Step: From Rejected Loan to Homeowner

I have known several people whose home-buying dreams have been threatened by poor credit. A poor credit score can mean anything from rejection of the loan application to a higher rate that makes monthly payments onerous.

Let's take a journey from having a poor credit score to being approved for a home loan, focusing on the step-by-step process facilitated by AI tools.

1. Initial Credit Score Analysis

There was a time when just viewing your credit score counted as a "credit check" and would have a negative mark on your credit. That is no longer the case, and there are numerous sources to instantly find your credit rating, such as your bank, websites, or apps dedicated to tracking and improving your score. We'll focus on the latter since, like with finance management in general, having an AI-powered app on your mobile device is the best practice for improving and maintaining good credit.

Some personal finance apps, like those mentioned in the previous chapter, offer credit-tracking services. However, these services often involve an add-on subscription. While having everything in one app can be beneficial, I'll focus on stand-alone credit improvement apps in this chapter. The **Tools and Tips** section has some worth exploring.

Upon signing up for an app and providing enough personal information for the credit report to be run (not more than a few questions), you will get an immediate view of your score and the items impacting it.

Scores below 700 can begin to affect both approvals and rates, with the challenges ramping up as you descend into the 600's. A credit rating of **620** is a standard cut-off for lenders even to consider a loan.

2. Factors Impacting Credit

When you receive a credit score, you should also get an overview of what has affected it. Here are some common factors:

- **High Credit Utilization:** Having high balances on credit cards, even if you pay them off each month.
- **Revolving Credit:** Not fully paying off balances in full each month, even if you are on time with minimum payments, has an adverse effect.
- **Late Payments:** These hurt your score and can have a lasting negative impact as they can remain on your record report for multiple years. Don't let it happen! At least pay the minimum on time.
- **Derogatory Marks:** Debt collections, bankruptcies, tax liens, or other significant legal actions.
- **Official Inquiries** (such as loan applications), **Credit Age, Number of Loan Accounts** (e.g., credit cards, car loans, home loans, etc.), and additional factors can have positive or negative consequences.

As you can see, there are numerous things to keep track of and manage.

3. Tailoring an Action Plan

The next step is to connect your financial accounts so that AI can evaluate your payment and credit usage histories, identify where problems might occur, and suggest improvements.

If you are at a **620** rating or generally in the 600s, you'll have several areas to attack. Here's what a typical analysis and plan might look like.

- **Reduce Credit Utilization:** The app will recommend a strategic plan to pay down the balances below a thirty percent threshold of your allowance on any single card or jointly. It will look at the card with the highest interest rate and suggest starting there. It will also show you how your payments can be made over a month to keep balances in check and eliminate revolving credit.
- **Increase Credit:** Open new cards to spread your credit needs further so you can keep any balance below thirty percent. Of course, this means more management effort with increased cards, so ongoing monitoring by AI will be necessary.
- **Late Payments:** The app will highlight the importance of setting up automatic payments for at least the minimum due to avoid future late payments. It can suggest contacting creditors to see if they would consider removing the late payment records in exchange for setting up automatic payments. Late payments can stay on record for years.
- **Disputing Errors**: AI might identify potential errors with debt that has already been paid off but is still listed as active. It is not uncommon for past issues to linger even once they are resolved. Apps will provide step-by-step guides on how to dispute errors with the credit bureaus and the path most

likely to succeed in clearing these, with time-saving methods.

- **Credit Building:** Responsible credit use and regular payments improve credit ratings. Certain apps provide credit-building options for those who lack a credit history or need to improve it. Regular reports are directly reported to credit rating agencies, demonstrating proof of successful loan management.

These are just some of the many things that could be identified and rectified. Reducing credit card debt may take a long time, so include a comprehensive financial management plan to reduce your monthly expenses and spending behavior.

4. Progress Monitoring and Adjustments

AI-based credit management is more than just a one-time assessment and plan. It can provide ongoing benefits such as:

- **Continuous Monitoring:** The app sends monthly updates on credit scores and any changes to the credit report.
- **Adjusting the Plan:** As credit card debt and other issues are addressed, the app updates payment strategies to optimize utilization across different accounts.
- **Fraud Detection:** Jumping on wrong charges that can absorb your credit allowance or become a permanent mark on your record ensures that all your hard work getting a good rating remains intact.

- **Educational Resources:** Throughout the process, personalized tips and articles on maintaining financial health will further empower informed decisions.

5. Achieving Results

If you have struggled with poor credit ratings, it may feel like a steep and even futile climb to improvement. Don't be discouraged! Credit ratings are constantly updated and can move significantly every month. It's very possible to increase your score within six months by a hundred points or more.

Once your rating climbs into the 700s, many more doors can open, including:

- Home, car, and personal loan approvals
- Eligibility for competitive rates
- Better credit card options

Credit checks and an improvement plan alone do not require AI, but knowing what to focus on, closely tracking the impact, juggling various loan sources, and determining the high-impact actions is AI's domain, as it utilizes troves of historical data from across the industry to provide recommendations.

The Power of AI Beyond Credit Scores

Please refer to the previous chapter on Personal Finance for the full range of how AI supports your financial health.

Savings

Money

A strong credit rating not only allows one to make life-changing purchases like a home or vehicle, but it can also save massive amounts in interest payments.

- **Lower Interest Rates on Mortgage:** Improving a credit score from 620 to 720 can reduce mortgage interest rates by at least one percent, if not more.
    - On a $300,000, thirty-year fixed mortgage, just a one percent reduction saves approximately *$60,000 over the life of the loan,* and a two percent reduction can save over twice that.
- **Avoidance of High Fees:** Hiring a financial advisor or a credit repair agency to analyze credit reports and suggest improvements can be costly.
    - Paying credit repair services or financial consultants deployed to handle these matters can cost up to *$2,000.*
- **Increased Credit Ceiling**: Good to Excellent Credit Ratings can increase how much credit you can access.
    - This ranges from *thousands to tens of thousands* in credit card limits to *hundreds of thousands* when trying to attain a home loan.

Estimated Net Money Savings: The amount good credit can save in lower interest rates and access to loans is counted in the *hundreds of thousands of dollars* over a lifetime. That's right, ***hundreds of thousands of dollars.***

Time

- **Rapid Analysis and Plan Generation:** Consulting with financial advisors could take days or weeks to understand a credit report and devise a plan to improve one's credit score. AI reduces this to a matter of minutes.
    - Approximately *ten to fifteen hours* of personal research and consultations.
- **Automated Monitoring and Updates:** Manual monitoring of credit scores and reports for changes and inaccuracies is time-consuming. AI automates this process, tracking your credit status and notifying you of updates.
    - *Two to three hours per month* spent on manual tracking and analysis and avoiding potential hours of trying to resolve mistakes or issues.

Estimated Net Time Savings: *Over twenty-five hours annually.*

Further Benefits

- **Increased financial confidence and peace of mind:** A good credit rating—or the path to getting one—will make you feel more in control of your financial situation, reducing stress and increasing your confidence to handle money matters.
- **Access to better financial products:** With a higher credit score, you become eligible for credit cards with better rewards, lower-interest personal loans, personal business loans, and other financial

products that were previously out of reach. Poor credit scores can block you from a promising future and life-changing opportunities, such as investing in real estate or starting a small business.

Tools and Tips

- **Credit Karma** is a comprehensive financial tracking tool incorporating free credit monitoring, reporting, and recommendations.[18]
- **Credit Sesame** targets credit racking and improvement as its central objective.[19]
- **Experion Boost** includes a mix of free services, offers, and a premium tier that can improve your scores and save you money.[20]
- **Self** and **Kickoff** focus on Credit Building, which requires a monthly "payment" commitment reported to credit bureaus. The money ultimately comes back to you, making it low risk, but it remains locked for a period of time.[21]

Look out! Many of these apps offer at least partially free services. Of course, nothing is free! These companies make money by promoting you to credit cards, loan offers, insurance, and other "financial services." While acquiring more credit can be a strategy for improving your credit

---

[18] https://www.creditkarma.com

[19] https://www.creditsesame.com

[20] https://www.experian.com

[21] https://www.self.inc; https://www.trainwithkickoff.com

rating, be attentive when signing up for cards and accounts that can lead to more debt.

A Partner to Your Dreams

Credit scores can seem intimidating, mysterious, and frustrating, especially since they act as a gate to critically important financial flexibility. With AI's help, your credit score can become your ally in accessing and achieving your most ambitious objectives, unlocking the ability to get the capital needed to pursue your dreams in education, business, or property. With the right AI-powered tool to improve and maintain a high rating, life's possibilities become unlimited.

CHAPTER 9

CUTTING COSTS

One of the benefits of connecting your financial accounts to AI-powered apps is discovering costs of which you weren't even aware. This turns into "easy money" that can immediately be saved and reallocated to more important (or enjoyable) things.

Many finance apps will do this as part of their services, while others may add an additional charge. Stand-alone apps dedicated to this specific activity are also available and should be at a lower cost than comprehensive finance apps.

Surprise, Surprise

It's time for me to get personal. While researching how AI analyzes and detects unwanted costs on your bank and credit card statements, I connected several apps to my accounts. I was looking to see how effectively they identified and categorized costs. I was not expecting to find errant or unknown charges. Was I wrong!

A recurring charges check is a simple and quick way to reduce costs and not waste money. In my case, here's what I discovered:

- A newspaper subscription and a magazine subscription I thought I had canceled.
- An entertainment club subscription I had no idea about!
- Monthly renter's insurance required for my son's temporary apartment—from last summer, six months ago!

These alone saved me over $75 a month. In addition, the apps can often cancel subscriptions for you. In my case, the entertainment club subscription required a phone call, but the AI surfaced the exact correct number to call.

Additionally, I decided to eliminate one of the streaming services I rarely used and even downgraded my TV cable subscription since that, too, was seldom used. I had gone for a higher tier once when they ran a special but forgot to reduce it when it was back to full price. And yes, that was years ago!

This added another $40 per month, saving me over $1000 per year without sacrificing anything I cared about! That cost savings could cover all the subscriptions mentioned in this book and save you a small fortune.

Step-by-Step: Cutting Unwanted Expenses

AI Tool: Rocket Money

1. **Select an App**

   As with other efforts in the finance category, the first step is utilizing an AI-powered app to scrub your expenses and flag ones that are potentially unused, redundant, or erroneous.

2. **Set Up Profile and Connections**

   Once you set up a profile and connect your financial accounts (usually about a two-minute job), the AI goes to work.

3. **Review List of Recurring Expenses**

   Here are the types of items it will identify, categorize, and even flag if recognized as a service or subscription people generally don't want or that seems suspicious.

   A. **Streaming services** include video streaming (Netflix, Hulu, Amazon Prime Video), music streaming (Spotify, Apple Music), and even live TV streaming services (YouTube TV, Sling TV). It's common to subscribe to multiple services; seeing them all together may make you wonder if this makes sense and motivate you to cut some!

   B. **Software and apps** encompass productivity tools (Microsoft Office 365, Adobe Creative Cloud), storage (Dropbox, Google Drive), and even specialized apps like photo editing or fitness. Software used to be a one-time purchase but is now usually a subscription—make sure you aren't paying for features or apps you rarely use.

C. **Magazines and newspapers** the AI identifies might be digital subscriptions to news outlets, magazines, or academic journals. It's worth evaluating which subscriptions you regularly read and benefit from and those you don't.

D. **Gym and fitness memberships** are easy to forget about. Given our often periodic enthusiasm for fitness, ensure you aren't paying for memberships you seldom use.

E. **Meal delivery services** like kit delivery services (HelloFresh, Blue Apron) or premium memberships for food delivery apps (Uber Eats, DoorDash) add convenience but at a high monthly cost if not fully utilized.

F. **Educational platforms** can be tricky. We often enroll in online courses and learning platforms (Coursera, MasterClass, LinkedIn Learning) for a class we've always wanted but then forget we are paying for it monthly.

G. **Gaming services** include online gaming platforms (Xbox Live, PlayStation Plus), game streaming services, or game subscriptions.

H. **Beauty and lifestyle boxes** are fun, but always confirm that subscription boxes for beauty products, clothing, or pet supplies (such as Birchbox, Stitch Fix, and BarkBox) are worthwhile.

I. **VPN services** are commonly needed for online access and privacy in certain countries when you travel, but they can be forgotten when no longer necessary.

J. **Charitable donations** can be made through automated services. Make sure these recurring charges to charities, NGOs, or Patreon support are what you are expecting.

By closely examining these and other subscriptions in a simple, organized format, you can identify which services truly add value to your life and which might drain your finances without a corresponding benefit.

Savings

Money

- **Reducing Unnecessary Subscriptions**: On average, individuals can save by canceling unused or redundant subscriptions.
    - **Savings**: *$20 to $30 per month.*
- **Optimizing Spending**: By analyzing spending patterns, AI can help you identify areas where you can cut back on non-essential expenditures.
    - **Savings:** *$50 to $100 per month.*
- **Avoiding Overdraft Fees**: By monitoring account balances and predicting upcoming bills, AI can help you avoid overdraft fees.
    - **Savings:** *$25 to $35 per incident.*

- **Detecting Fraud Early**: Quickly detecting fraudulent transactions can prevent more significant losses.
    - **Savings:** *You can save hundreds or thousands of dollars* by quick action initiated by AI.

Estimated Net Money Savings: *$100 to $200 monthly*, not including the prevention of more severe fraud.

Time

- **Automated Analysis:** Traditional methods of financial tracking involve manually reviewing statements, categorizing expenses, and identifying trends.
    - **Savings:** *Several hours each month.*

- **Subscription Management:** Identifying and canceling unused subscriptions can be time-consuming and involve multiple steps to confirm cancellation. Many AI-driven apps streamline this process, either canceling them with a confirmation button or providing a direct number to call.
- **Real-Time Alerts**: AI systems can notify you immediately of any unusual spending or duplicate charges, eliminating the need to check accounts periodically for errors or fraud. This saves users from the often tedious task of fraud detection and dispute resolution.

Estimated Net Time Savings: *Five to ten hours per month.*

Tips and Tools

- **Rocket Money** and the aptly named **Cancel Subscriptions** specialize in identifying and canceling unwanted subscriptions and recurring costs.[22]
- More comprehensive personal finance apps include recurring expense analysis in their services, although some charge extra for managing and canceling subscriptions.
- While an ongoing subscription is generally the best way to ensure you get the most benefits from apps helping you manage your finances, if you want to scrub your monthly expenses for unwanted subscriptions and recurring costs, the free trial period will provide plenty of time to accomplish this objective. At a minimum, sign up for the free trial, eliminate those unneeded costs, and then cancel.

Easy Money

By identifying wasteful expenditures, flagging unusual costs, and providing actionable insights, AI empowers you to use every dollar you spend towards your preferred outcome. Many money-saving efforts are available without the new technology but require regular attention to detail. Even if you are diligent about reviewing monthly statements, you may be surprised how easy it is to overlook costs in long lists of poorly labeled credit card and bank statement transactions. I certainly was!

---

[22] https://www.rocketmoney.com; https://www.joinchargeback.com

CHAPTER 10

SLASH YOUR ENERGY BILL

In the quest to manage household expenses, energy bills often stand out as a significant monthly outlay. The average energy cost in the U.S. annually is over $2000. With less predictable and more extreme weather patterns expected in the coming decades, it's become more challenging to manage energy at the grid level, and you will end up paying for inefficiencies.

You have likely heard about the smart home, and maybe you've even added some elements, such as sensors for your lights. The most significant and accessible action you can take regarding your energy costs starts with your thermostat. Based on the price of energy and the latest and greatest devices, you'll, at worst, break even in year one, and then it's savings, savings, savings from there on out. Not only are you cutting costs, but you'll be contributing to a healthier planet.

Step-by-Step: Raise Your Home's Intelligence

1. Find a Smart Thermostat

There are several smart thermostat choices, and more will enter the market every year. What you want to look for are thermostats that:

a. Learn how to adjust based on your living patterns—this is AI at its best!

b. Either use sensors to detect when people are in the house, or even better, use the geolocation on your phones so you don't need the extra cost of sensors!

c. Connect and communicate through an app so you have real-time control and information no matter where you are in the world.

2. Installation

Smart thermostat manufacturers have made it very easy to install. There will usually be wired and wireless options—both work equally well, but you need to be sure to change batteries on the wireless as required. In this regard, wired is a better option in the long run, saving on the cost of batteries and reducing environmental impact.

Are you stuck with the manufacturer's instructions? As we saw in the DIY chapter, AI can assist in installing almost anything in your home with prompts or even a picture of your installation area.

3. **Set Up and Go**

   The smart thermostat applications will walk you through a short setup. They will initially use your answers to critical questions, such as time around home, room use, preferred settings, etc., to program a plan. Some even tap into weather trends to optimize based on historical and real-time data.

Once the thermostat is running, AI will learn about your patterns and preferences, the temperatures inside and outside your home, optimal times to heat and cool the house, and many other aspects. If you have an Energy Sensor on your panel or a way to connect to your power company's metering wirelessly, it can factor in when energy costs are lowest.

What AI Can Do and We Cannot

How smart thermostats work is an excellent example of something that no matter how hard we try, we are simply not going to be able to do as well as AI, and certainly nowhere near as quickly and efficiently.

Let's take the example of a three-bedroom house in any suburb in the country. Each home will have several factors that impact what it needs to heat or cool efficiently. These include:

1. Daily and hourly weather
2. Amount of windows and positioning of them toward the sun over a day
3. Retention of temperature based on house insulation and building materials
4. Sealing, weatherstripping, and window glazing
5. Reflective roofing materials

As you can imagine, attempting to understand how each of these factors affects temperature control in your home and potentially investing in improvements or replacements could be immensely time-consuming and costly—probably hundreds of hours and thousands, if not tens of thousands, of dollars.

A smart thermostat can't change your house's physical conditions or weather pattern. But it can adjust instantly and intelligently, adapting to your home's temperatures, personal schedule, and air conditions.

As we will see in the Savings section, the result is thousands of dollars over the life of a typical home ownership.

How AI Can Save You Even More

While heating and cooling account for about fifty percent of the typical energy expense, other smart home additions can further enable AI to help you save on costs and prevent the inefficient use of natural resources.

1. **Energy Sensors**
   - These connect to your electrical panel and analyze energy usage patterns across your complete system, including lights, appliances, and HVAC. Tracked through a wirelessly connected app providing AI-driven insights, they can lead to the optimal times to run heavy appliances, provide data on where lighting or other energy drains are occurring, and recommend optimizations.
2. **Smart Lighting**
   - These generally involve smart bulbs tied into an app and don't require new lighting fixtures.

AI can recommend schedules for when lights should turn on or off based on daily routines or even on your location, as identified via your phone's GPS. It will analyze usage over time and suggest or auto-implement optimizations to save energy, such as adjusting brightness levels or changing color temperatures.

3. **Water Usage**
   - Smart water meters can predict your water consumption patterns and detect leaks early, preventing wastage and reducing water bills.

Savings

Money

While every household is different, the leading smart thermostats have demonstrated they can deliver a minimum twenty-five percent reduction on your energy bill. In places with cold winters and hot summers, these bills can exceed $2000 per year, thus providing more than *$500 per year in savings.*

Net Money Saved: Top-of-the-line smart thermostats cost around $250. Even if you paid an electrician to install one, you would break even in year one and save continuously in the future. Throughout average home ownership (five to seven years) this translates to *thousands of dollars.*

Time

Most of us don't spend much time adjusting and re-adjusting our heating and cooling. We also tend to overuse our systems, leaving them on when we're not home or forgetting to adjust them when we leave a room. Sure, we have timers on them, but that is a very loose way to manage a costly resource.

The fact that we don't invest the time into adequately managing our thermostat is another way to see how AI augments our lives. The reality is that trying to adjust our thermostat every day for the numerous variables that impact air temperature is not even possible. AI can do something we couldn't accomplish even with time and effort.

Tools and Tips

- **Ecobee** and **Tado** are two companies worth looking at for their smart home products, including thermostats.[23]
- **Google Nest®** and **Honeywell** also produce smart thermostats. It's a good idea to read reviews before purchasing.[24]
- Because AI's accelerating power is relatively new to the market, hardware devices may take longer to catch up. Even with firmware updates, leveraging the vast possibilities AI unlocks will require building with it in mind from the ground up.

---

[23] https://www.ecobee.com; https://www.tado.com

[24] https://store.google.com; https://www.honeywell.com

- **Consumer Reports**, **Wirecutter**, and **Wired Magazine** all run independent and objective tests of electronic hardware like smart home devices and are good sources to find reviews.[25]

A Clever House

AI offers a practical and effective solution for managing and reducing household expenses, particularly energy bills. By leveraging its ability to analyze and predict usage patterns, you can enjoy substantial savings and contribute to environmental conservation.

These capabilities underscore AI's role as a technological marvel and a day-to-day utility that empowers you to lead more efficient, cost-effective, and environmentally friendly lives. The benefits affect not only you but your community and the future of everyone on the planet.

[25] http://www.consumerreports.org; http://www.nytimes.com/wirecutter; http://www.wired.com

SECTION IV

PRODUCTIVITY

CHAPTER 11

BUILDING YOUR CAREER

A resume is not just a list of accomplishments but a reflection of your professional identity. It is also usually the first impression employers have of you. However, tailoring your resume for each job application can be time-consuming and daunting.

AI transforms the resume-building process into a streamlined, personalized experience. Let's see how it can assist in creating a dynamic resume, focusing on customizing it swiftly for a specific job application.

Master a Resume with AI

Most resume-building platforms, if not all, have incorporated AI into the process. There is still quite a difference between them, and some of the newer ones built using AI from the ground up are stronger in their application.

For this example, we'll use **Rezi**.[26] However, as always, it's essential to check out a few options to find the right one for you. Most will allow you to get a feel for the tools, if not create a complete resume, for free before prompting for a subscription fee.

Rezi is good to experiment with as it allows a thorough experience with an AI-powered resume builder. It helps streamline the job application process, creating resumes optimized for the commonly used Applicant Tracking Systems (ATS) and enhancing your chances of catching a recruiter's eye. Most companies and recruiters use ATS to do an initial scan of a resume to see if you're a fit for the position. The sad truth is that a human being never sees most resumes submitted!

Step-by-Step: Building a Resume

AI Tool: Rezi

1. Create an Account and Start a New Resume

2. Input Personal Information

- Basic information such as name, contact details, and links to professional profiles (e.g., LinkedIn).
- Linking profiles is helpful as it will use this information to make recommendations to personalize your resume.

[26] https://www.rezi.ai

3. **Add Your Work Experience**
   - Use the 'Add Experience' section to list your previous jobs.
   - Input your role, the company name, and the duration of your employment.
   - Describe your responsibilities and achievements. AI will suggest action verbs and job-specific phrases to enhance your descriptions.

4. **Include Education and Certifications**
   - Highlighting educational achievements pertinent to your career field.

5. **Feature Skills and Languages**
   - The AI recommends skills based on your experience and desired job field.

6. **Finalize the Resume Layout**
   - Choose from various professional templates that suit your style.
   - Easily customize the layout by adjusting fonts, colors, and section order.

7. **Review and Optimize**
   - Use real-time content analysis to receive feedback on common errors and optimization opportunities.
   - AI will ensure the resume is ATS-friendly.

8. Download the Resume

- Once satisfied, download your resume in PDF format, ready to be submitted to potential employers.

Step-by-Step: Customizing for a Specific Job

Imagine you've found your dream job listing. The position matches your skills and experiences perfectly, as do the qualifications of hundreds of other applicants. To stand out, you must customize your resume to highlight how your background aligns with this job.

The strength of AI-powered platforms like Rezi is their use of advanced algorithms to tailor content specifically for a job you're applying for. You provide a specific job description, such as Marketing Manager, and it suggests changes and additions to make the resume more suitable for particular roles. AI even recommends the layout that best matches the industry standards. This level of customization, which could take hours manually, is achievable in minutes. Here's how that would work.

1. Job Description Analysis

- Input the job description into Rezi.
- The AI analyzes the description to identify critical skills and keywords that should be included in your resume.

2. Tailor Your Work Experience

- AI suggests modifying your work experience entries and highlighting specific marketing skills and achievements that align with the job description.

- Incorporating relevant marketing terminology and action verbs demonstrates your field proficiency—AI does it for you!

3. Emphasize Relevant Skills

- Adjust the skills section to prioritize marketing-related skills.
- Include any software or tools listed in the job description that you are proficient in, such as CRM software or digital analytics tools.

4. Update Your Objective or Summary

- Assure your resume summary or objective reflects your career goals and enthusiasm for the role of a marketing manager.
- Ensure it contains keywords from the job description to improve visibility in ATS scans.

5. Customize the Design

- Choose a template that fits the culture and professionalism of the marketing industry.
- Ensure the design remains clean and the text easily scannable.

6. Optimize and Review

- The final review feature ensures that your customized resume matches the job description as closely as possible and provides additional tips to improve its impact.

7. Download and Apply!

An AI-powered Resume Builder like Rezi is ideal for applying to multiple positions, as each application should be customized. You save time and enhance the quality of your resume, ensuring it is aligned with industry standards and job-specific requirements. Whether you are a recent graduate or a seasoned professional, AI can improve your chances of getting through the door and being interviewed.

Are you applying for a really important position? Visit another site like Jobscan, which is optimal for checking already-created resumes and provides several scans in its free tier. You can also get some further tips to make it even better.

More Resumes, More AI

Resume builders offer a wide range of services to enhance your job application process. AI can assist in the following ways:

1. **Resume Customization**
   - Analyze job descriptions and tailor your resume to match, emphasizing the skills and experiences most relevant to the position.
2. **ATS Optimization**
   - Optimize your resume for ATS, which employers use to filter resumes. By identifying and using key phrases and formats, AI can ensure yours passes through.

3. **Layout and Design Suggestions**
    - Automatically suggest and apply the most effective layout and design based on your industry and the specific job, making your resume visually appealing and readable.
4. **Skill and Gap Analysis**
    - Identify any potential gaps in your resume compared to industry standards or specific job requirements, suggesting areas for improvement or highlighting.
5. **Performance Tracking**
    - Some AI tools even offer the ability to track how well your resume performs in job applications, providing insights for further refinement.

Savings

Money

- **Resume Writing Assistance:** Not so long ago, you would have had to pay a resume writer up to $500 to build and optimize a resume, and then that would still need to be customized for each position. Now, you can get quite far at no cost, and even if you invest in one to two months of subscription to an AI-powered site, you will receive weeks of submissions with many personalized resumes tailored to each position. Monthly subscription prices range from $10 to $30.
    - **Savings:** *hundreds of dollars,* and the money you spend comes with far greater value.

Time

- **Tailoring Resumes to Jobs:** Crafting a customized resume for each job application can take hours. AI shortens this time, allowing you to submit many more applications faster.
    - **Savings:** *Hours per job application.*
- **Building a Resume:** Creating an original resume used to take hours across multiple days. An original resume takes about sixty minutes with AI tools.
    - **Savings:** *Hours* with higher quality results.

Estimated Net Time Saved: This actual count may be in hours, but the speed and flexibility of resume creation and customization really save days. Applying to new jobs quickly will accelerate your entire process, allowing you to invest much more time in finding great jobs to apply to.

Opportunity

- **Opening Doors**: By ensuring your resume is closely aligned with job descriptions and optimized for ATS, AI increases your chances of reaching the interview stage. This optimization can be the difference between landing your dream job or being overlooked and never getting the shot you deserve.

Tools and Tips

As previously noted, AI has become a significant factor in resume-building platforms, and you will find a lot of choices! Below are some sites that specialize in AI and could be worth looking into, but there undoubtedly are many other existing and new sites incubating, so when this becomes a priority, spend a little time upfront finding the right platform. As always, investing in a few during a free trial or low-cost entry period can be very helpful. Your career is worth it!

- **Rezi.AI** is designed to streamline the job application process by providing tailored resume enhancements and optimizations to increase applicants' chances of passing through ATS.
- **ResumeGenius** creates your resume in minutes, with templates and examples.[27]
- **Jobscan** uses AI to compare your resume against job descriptions, providing scores and insights to improve your match rate.[28]
- **Zety** provides AI-powered advice on content and layout, emphasizing making your design eye-catching and appropriate to the field.[29]

It's good practice to regularly update your resume with AI tools, even when not actively job searching, to keep it ready and optimized for future opportunities.

The Ultimate AI Benefit: Career Opportunity

The journey to your next job doesn't have to start with the daunting task of resume building. AI not only simplifies the process but also significantly enhances the quality and

[27] https://www.resumegenius.com

[28] https://www.jobscan.co

[29] https://www.zety.com

relevance of your document. For years, recruiters have used tools to eliminate resumes without even looking at them. AI can be your tool to counter and even excel in these common hiring systems.

By leveraging AI to customize your resume for specific job applications, you're not just applying; you're presenting a tailored narrative of your professional journey, dramatically increasing your visibility and appeal to potential employers.

CHAPTER 12

LANGUAGE TRANSLATION

In our increasingly interconnected world, the ability to communicate across linguistic boundaries has never been more critical. Language translation, once a task for skilled human linguists, has been transformed by AI. AI-powered translation breaks down language barriers with remarkable efficiency and accuracy.

The revolution in translation enabled by AI has multiple aspects. We are already benefiting from accurate captioning in any language in video, and we're incredibly close to real-time translation in speech. Many of you may have already seen how anyone speaking on video can be translated as if they are speaking a language they don't know, with the lip sync incredibly accurate.

While all these advancements will majorly impact the future, let's look at one aspect that gives us a new lens into the past.

Reveal Family History

Imagine discovering a box of old, handwritten letters in your attic, penned by your great-grandparents in an old-world language such as Yiddish. These letters could hold family stories, historical insights, and emotional connections to your past. However, the language barrier might seem insurmountable if you don't speak it—and, in likelihood, no one in your family or community speaks it either.

Enter AI-powered translation tools that decipher handwritten texts, translate them into English, and unlock the stories concealed within those pages. This technology not only translates but preserves the essence and nuances of the original language, providing a bridge to the past for future generations.

Incredible Stories

All of this can be done by taking a simple photo of the letters on a smartphone or scanning them on a computer and then uploading them for translation to ChatGPT. ChatGPT's latest LLM model outperforms every translation and Optical Character Recognition (OCR) tool to date, including the venerable Google Translate.

Step-by-Step: Decoding Secret Letters

In one such case in my own family, handwritten letters were translated, revealing never-before-known correspondences between relatives in South Africa and Lithuania in the late nineteenth and early twentieth centuries.

AI Tool: ChatGPT

1. **Take a Picture or Scan the Document**

   Current smartphones take high-resolution images that provide plenty of detail.

2. **Use Attachment Feature in Chatbot Text Box**

   Add the image(s) and prompt the chatbot to translate it into English (or whatever language you choose). Include any information you have about the documents, such as the language, time period, and country of origin.

3. **Get the Translation**

   Yup, it's that easy!

4. **Ask About Specific Words or Phrase**

   Use a screenshot tool to isolate specific words or phrases and prompt the Chatbot for more information. Maybe it's the name of a place, a historical person, or a word or phrase that didn't seem to translate correctly. Provide context from family history to help the AI understand the true meaning of anyone's words.

Translating personal letters is far more challenging than books or more formal manuscripts. Letters are filled with colloquialisms, local dialects, and personal intimations that require a sophisticated comprehension of language, culture, and customs.

Yet the AI could recognize and decipher the old Yiddish of these correspondences, a language that is rarely used today.

Revelations

In this case, the communications between relatives across continents revealed how money was sent to support oppressed family members in a poor rural village, circumventing efforts by the countries to block such efforts. In one case, the money enabled the purchase of a business to help the family escape poverty. In another case, the money went to enabling other relatives to leave the Eastern European village before the death and destruction of World War II overran it.

The history of these families is now preserved for future generations. Stories of escape, as well as the tragic tales of those who did lose their lives in the Holocaust, were preserved in these letters.

What AI Can Do in Language Translation

AI-powered translation has evolved to offer an array of services that cater to various needs.

Its features include:

1. **Text Translation**
   - Instantly translates written content from one language to another.
2. **Speech-to-Text Translation**
   - Converts spoken words into written text in another language.
3. **Real-Time Conversation Translation**
   - Offers immediate translation during live conversations, breaking down communication barriers in real-time.

4. **Handwritten Text Recognition and Translation**
    - Deciphers handwritten notes, letters, or documents and translates them into the desired language.
5. **Cultural and Contextual Nuance Recognition**
    - AI algorithms can now understand and interpret cultural references and context, ensuring translations are literal but also appropriate and meaningful.

Savings

Before AI, translating a document required a human translator, costing both time and money. AI translation can accomplish the same task in seconds, usually for free. This efficiency allows individuals, businesses, and researchers to access information and communicate across languages without significant investments.

Money

- **Translation Services:** Professional translation services typically charge between ten and thirty cents per word. Due to the rarity of specialized languages like Yiddish, rates can be higher, ranging from twenty to fifty cents per word.
    - **Savings:** *$800 or more* for four letters of 1000 words. Every additional letter would incur more fees. The example had ten letters, so the cost would have been sizably more.

Time

- **Translation:** Finding a specialist, sending the documents, and waiting for the translations usually takes weeks.
    - **Savings:** Translations are now available in *seconds,* no matter where you are in the world.

Heritage

The ability to translate old handwritten letters or documents can save a part of history that might otherwise be lost. By unlocking the stories and knowledge in these texts, AI helps preserve cultural heritage and personal history, enriching our understanding of the past and our place in it.

Tools and Tips

- **ChatGPT** translates text or images by inserting them in the prompt field. It instantly recognizes the language and translates it to the language of your request.[30]
- **Google Translate** offers text and website translation in multiple languages. Its camera feature can translate printed or handwritten text in real-time.[31]
- **Microsoft Translator** provides text, speech, and document translation in various languages, including real-time conversation translation.[32]

---

[30] https://chatgpt.com/

[31] https://translate.google.com/

[32] https://www.microsoft.com/en-us/translator/

- **MyHeritage In Color™** is not a translation tool, but it helps enhance old documents or photos and make handwritten texts clearer for accurate translation.[33]
- Use high-resolution images of handwritten texts when using translation apps to improve accuracy.

Bridging Cultures

The evolution of AI in language translation has opened a world of possibilities, from conducting international business effortlessly to unlocking personal histories in old letters. By harnessing AI, we can bridge cultural and linguistic divides, bringing the world closer together.

As AI advances, its potential to enhance our understanding and appreciation of diverse languages and cultures grows. This chapter has showcased how AI-powered translation can be a powerful tool in eliminating language barriers, allowing us to connect and learn from each other in ways we never thought possible.

[33] https://www.myheritage.com/incolor

CHAPTER 13

THE PERSONAL ASSISTANT

It is regularly said that AI will help humans tremendously. Usually, that refers to the large-scale impacts it will have on society, such as scientific discoveries, cures for diseases, or automation of our industrial and transportation systems.

Yet, day to day, what most of us need help with is managing intense schedules in an incredibly fast-moving world. Fortunately, AI offers innovative solutions to streamline our lives by providing us with what we all wish we had: a 24/7 personal assistant. Layering a chatbot on top of our tangle of communications, calendars, and task lists can be a daily lifesaver.

Conquer Daily Demands

Let's take my friend Emily. As a project manager at a bustling tech startup and a mother of two young children, her days were a delicate dance between work commitments and family time. Despite her best efforts, Emily often found herself overwhelmed, struggling to manage her daily obligations, to-do list, and schedule. It seemed she was always a step behind, racing from one place to another.

The Breaking Point

On a particularly chaotic Wednesday, Emily missed a vital work deadline and her daughter's ballet recital. Exhausted and disheartened, she sat down that evening, searching for a solution to regain control over her life.

The answer was incredibly close at hand. A simple voice command away, in fact.

Most of you are likely familiar with voice assistants such as Alexa, Google Assistant, or Siri. You may not realize how much AI has powered these historically finicky platforms with a much greater ability for voice recognition and integration across common calendaring and task list applications.

Enable notifications that are already common in these applications, and the 24/7 personal assistant has arrived.

Choosing a Platform

Emily turned to Google Assistant since G-Suite was already handling her workplace and personal calendar. Apple has similar applications, and Amazon's Alexa integrates with both of these as well. As always, look out for new entries into the space—they show up practically daily!

Step-by-Step: Your Personal Assistant

AI Tool: Google Assistant

1. Activate Google Assistant

- This comes natively on Android devices or can be downloaded on iOS-powered hardware such as an iPhone or MAC.
- Say "Hey Google" to activate the Google Assistant and walk through the setup.

2. Connect Calendar

- If you used a Google account to set up, your Google Calendar is automatically integrated and linked!
- You can check if the right calendar is linked and add other Google Calendars with other logins if needed. This is handy if you have separate work and personal accounts.
- Want to add an appointment on any day or time? Say, "Hey, Google," and tell it. For example, "Hey Google, add pick up Zoe at Dance at 3 p.m. today on my calendar and remind me thirty minutes in advance."
- If you need to find time for a work task such as writing a strategy memo, say, "Hey Google, when do I have one hour tomorrow free?" and then prompt it to add to your calendar with the title of your choice.

3. Add Reminders

- Separate from the calendar, reminders can be set at a specific time. For example, "Hey Google, remind me to call the dentist to make an appointment at 1 p.m. tomorrow."

4. Utilize for Task Management

- You can make or add to lists at any time. Just prompt, "Hey, Google, take a note" or "Create a list."
- Many non-Google task apps integrate with the Assistant, so you can likely use whatever you are most comfortable with or what your work requires.

5. Activate Notifications

- To receive push notifications, ensure your device's notification settings are enabled for Google Assistant or whichever voice assistant you use.

6. Review Your Day

- An excellent way to start each day is to ask your assistant for a review. For example, while you're getting dressed, say, "Hey Google, what's my day like?"
- Google Assistant will provide a rundown of scheduled tasks, meetings, and reminders.
- Missing something? Tell it to add it, as if talking with an actual assistant.
- If you are unsure when you have an open slot, say, "Hey, Google, when do I have an hour free?"

7. Customize Your Assistant

- Most voice assistants offer various voice options, including gender, accents, and attitude.

- Language options are also quite plentiful. Emily sometimes asked her assistant to provide her tasks in French, which she studied at university so that she could have some privacy with her kids around!

AI's Capabilities in Personal Management

There are many ways AI can help you manage the frenetic pace of the modern world.

AI tool capabilities include:

Scheduling

- Manage your calendar, including adding, updating, and canceling events.
- Coordinate with multiple participants' calendars to find the best meeting times.
- Schedule meetings across different time zones by automatically converting and displaying times accurately.
- Book appointments with third parties (e.g., doctors, hairdressers) by interacting with their scheduling systems.

Task Management

- Create and assign tasks based on voice commands or text input.
- Prioritize tasks based on deadlines, importance, and user preferences.
- Track the progress of tasks and provide updates or reminders.

Reminders

- Send reminders for upcoming meetings, deadlines, and events.
- Location alerts, such as reminding you to buy groceries when you're near a store.
- Set notifications for daily routines, such as medication schedules or exercise.
- Nudge you to follow up on emails, tasks, or meetings.

Communication

- Filter, prioritize, and even respond to emails.
- Screen calls, take messages, or route important calls to you.
- Transcribe and summarize meetings, highlighting key points and action items.

Personalization

- Learn your preferences and habits to provide more personalized assistance.
- Understand the context of your requests and provide more relevant assistance.
- Improve its performance over time by learning from your interactions and feedback.

Integration

- Integrate with productivity apps (e.g., Trello, Asana, Slack) to streamline workflows.
- Connect with smart home devices to manage your home environment, like adjusting thermostats or turning off lights based on your schedule.

And More!

On any given day, feel free to give these prompts a whirl with your assistant:

- Weather updates and dressing suggestions.
- Traffic updates and optimal route suggestions.
- Quick research and information retrieval.
- Health check-in: how many steps did you do today, or how far did you walk?

Savings

Money

While the primary savings from an AI personal assistant will be in time and sanity, there can be monetary savings:

- **Virtual Assistants:** The market has become commonplace for many and is now a fifteen-billion-dollar industry expected to double, if not quadruple, over the next few years with the addition of "professional" AI Voice Assistant services. Using such a service can cost you or your employer hundreds to thousands of dollars per month.

- **Savings:** Whatever the cost, generally *thousands of dollars* can be saved as much of this can now be done for free with AI tools.

Time

- **Time Management**: Getting assistance organizing your everyday life can save you time. This will vary greatly depending on your situation.
    - **Savings:** *Thirty minutes per day, or almost two hundred hours annually,* with an AI-integrated solution efficiently managing calendars, setting reminders, and planning tasks. That's four working weeks' worth of time!

Sanity

- **Stability and Control**: This must be one of the most significant savings in the book! Helping manage your day-to-day life and providing just-in-time reminders could save your job, family, heart, and mind. No one can put a price on that!

Tips and Tools

- Voice assistants such as **Siri**, **Google Assistant**, and **Alexa** can offer amazing free help in your daily battle with time, hectic schedules, and endless to-dos.

- For more advanced users, integrating applications like **Todoist** and **Trello** with your AI Voice Assistant can add sophistication to your task management support.[34]
- Most voice assistants can integrate with any major calendar applications—Microsoft Outlook, Google Calendar, or Apple Calendar—so you're not required to remain within one specific platform provider.

A Pocket Aide That Never Sleeps

Leveraging AI to manage your daily schedule, reminders, and tasks enhances productivity and ensures a well-balanced lifestyle. By customizing a personal AI assistant with your applications and preferences, you can easily navigate your day, gaining more productivity in work and more joy with personal and family time that is not rushed or missed.

[34] https://todoist.com; https://trello.com

CHAPTER 14

INSTANT ANSWERS TO ANYTHING

The internet has provided a tantalizing abundance of information, with boundless knowledge at our fingertips. The quest for answers often leads us through a maze of sources, each with an agenda not always clear or concerned with the truth. "One click away" generally provides a promoted answer paid for by someone vying for our attention. Those of us determined to extract the good info from the bad often conclude our search after countless clicks and many hours. By then, we're confused and exhausted.

Powered by advanced AI algorithms, chatbots represent a beacon of clarity in this chaos. They can provide immediate, accurate responses to a myriad of questions, ranging from the simple to the complex. They don't represent just one source but the entirety of the internet.

This has obvious uses, from history to geography, math, grammar, or people and culture. This could be an endless chapter on infinite questions, but I'll leave the fun of discovering the vast knowledge of AI chatbots to your curiosity and exploration.

Navigate Personal Uncertainty

While the depth of knowledge is impressive, one of the more intriguing uses of querying a chatbot is how it can act as an intelligent sounding board for more complex questions and decisions. Let's look at an example you might never have considered to consult your chatbot about: deciding about job relocation.

Imagine being at a crossroads, faced with a decision that could alter the course of your personal and professional life. The options seem equally daunting, and the fear of making the wrong choice leads to paralysis. You might find yourself in this situation if offered a job in another city, state, or even country, and you must decide if you should uproot your life, leaving friends, family, and many other things you covet.

Friends Try to Be Helpful, But...

A natural way to make such an important decision is to ask friends. If you have ever been in this situation, you'll know that almost everyone has an opinion, and they are not only at odds with one another but charged with emotional reactions or swayed by typical human biases. Whether it's the job, the location, or the personal relationship, it can be hard to navigate the advice. Even the most well-meaning friends can confuse and convolute your thinking.

What you need is an impassioned, brilliant, and knowledgeable advisor. And now we all have one! We even have a few with no cost and rapid responses.

A Trusted Confidante

With no one watching or judging, you can query the chatbot with all your ideas, anxieties, and questions. In the context of job relocation, you are free to express all the reasons you can think of to take it and all the ones to decline it. What's helpful about a chatbot is that it will not throw out opinions but walk you through the decision-making process. Here's how it might approach the job relocation question.

Step-by-Step: Relocate to Cleveland?

1. Ask the Chatbot

With no idea what to expect, I queried ChatGPT about a hypothetical move from Los Angeles to Cleveland. The chatbot walked me through six questions, which took about two minutes to complete. They centered on why I considered the move and its pros and cons.

The analysis that came back summed up the challenge quite well.

- Pros: a better paying job, career position, and opportunity to explore a new place.
- Cons: miss friends, the West Coast, and access to nightlife and entertainment

2. Deepen the Questioning

Over the next few minutes, the conversation became much more interesting as I probed into the cons. The chatbot proposed various ways to meet new people, samples of the cultural and recreational opportunities, and ways to get to know Cleveland on a pre-decision visit.

I asked about the dating scene compared to Los Angeles, and it offered a genuine analysis. It identified the challenges of LA, with its high cost, large travel area, and competition. Cleveland, it noted, had "a small community vibe" that was better to make meaningful connections. While it would have a smaller dating pool, there would also be less frenzy and competition and more manageable expectations. It also gave specifics on the cost of a date in Cleveland versus LA—wow, if you're dating, you'll save a lot of money in Cleveland!

I queried it: *As my friend, what do you hear about Cleveland?*

The chatbot provided a detailed list of pros and cons. On the negative side, it mentioned the often inclement weather and higher crime. I probed the latter and inquired about safe areas to live. It provided a detailed analysis of several neighborhoods with statistically much better crime rates and described the culture and dynamics of the regions.

Regarding affordability in general, I asked for specifics. See the Savings section for that response. There were no holding back details here!

3. A True Collaborator

In less than ten minutes, I had a load of information about such a decision. Compare that with the time and efficiency of a meeting or call with a friend to discuss the options. This would take one to two hours each. AI provided more precise, detailed information and answered questions that (realistically) no friend could offer—even one who had lived in Cleveland!

Don't get me wrong, please spend a lot of time with friends. You just might want to keep the conversation to other topics.

Cleveland sounds like a pretty good option, I told my chatbot.

It quickly replied: *Cleveland could be a great fit for you! With its lower cost of living, rich cultural scene, and a more community-oriented dating environment, it offers many benefits compared to Los Angeles. You might find the smaller, close-knit atmosphere refreshing, and there are plenty of opportunities to meet new people and engage in local activities. The affordability and ease of getting around could also make your transition smoother.*

Maybe it's right!

Ask AI Anything: Relocation Opportunity

As the chapter title asserts, you can ask AI anything. In our example on relocation, here are the many questions it can answer—and I am sure a million others you can think of! AI can help you with:

1. **Research and Information Gathering:**
    - Cost of Living Comparison: AI can provide detailed comparisons of living costs between

different cities, including housing, groceries, utilities, and transportation.
- Neighborhood Analysis: AI can analyze crime rates, school ratings, and amenities to identify the safest and most suitable neighborhoods.
- Job Market Evaluation: AI can assess job opportunities, industry presence, and average salaries in the new location.

2. **Personalization and Recommendations:**
   - Housing Search: AI-powered platforms can recommend homes or apartments based on your preferences and budget.
   - Lifestyle Matching: AI can suggest areas that match your lifestyle preferences, such as proximity to parks, cultural venues, or nightlife.
3. **Logistical Support:**
   - Moving Services: AI can help you find and compare moving companies, storage solutions, and other relocation services.
   - Travel and Commute Planning: AI can provide information on public transportation, traffic conditions, and commute times.
4. **Financial Planning:**
   - Budgeting Tools: AI can assist in creating a relocation budget, accounting for all potential expenses.
   - Tax Implications: AI can provide insights into the tax differences between your current location and the new one.

5. **Social Integration:**
   - Community Insights: AI can help you understand the social dynamics of the new area, including demographics and community activities.
   - Networking Opportunities: AI can suggest local events, groups, and networking opportunities to help you integrate into the new community.
6. **Personalized Advice:**
   - Decision Support: AI can use your preferences and priorities to generate personalized recommendations and insights to aid your decision-making process.

Savings

Money

The cost of living was one of AI's significant advantages in moving from Los Angeles to Cleveland. Here was the detail it gave in less than ten seconds to substantiate its claim:

- **Housing**
  - Cleveland: Median home price is around $363,504; median rent for a two-bedroom apartment is $1,254.
  - Los Angeles: The median home price is around $1,248,505, and the median rent for a two-bedroom apartment is $3,184. Housing costs can be 300 percent higher in Los Angeles.

- **Utilities and Transportation**
  - Cleveland: Utilities for a 915-square-foot apartment average $248.54 monthly. Gasoline costs around $3.36 per gallon.
  - Los Angeles: Utilities are slightly cheaper at $194.27, but gasoline is much higher at $5.34 per gallon.

- **Food and Groceries**
  - Cleveland: Generally more affordable; for example, a gallon of milk is $4.62, and a dozen eggs cost $3.97.
  - Los Angeles: Higher prices with a gallon of milk at $5.01 and a dozen eggs at $4.50.

- **Entertainment and Dining**
  - Cleveland: A meal at an inexpensive restaurant costs about $17, and a three-course meal for two at a mid-range restaurant costs $68.
  - Los Angeles: Dining out is more expensive, with an inexpensive meal costing $25 and a mid-range three-course meal for two at $120. Movie tickets are also higher at $19.39 compared to Cleveland's $12.21.

- **Healthcare**
  - Cleveland: Doctor's visits average $119.
  - Los Angeles: Slightly more expensive, with doctor visits at $130.

- **Childcare and Education**
  - Cleveland: Monthly preschool costs about $949.
  - Los Angeles: Significantly higher at around $1,950 per month.

Even if your friend who lived in Cleveland had been the mayor of the city, I don't think they would know all these details and be able to share them instantly.

Time

- **Research Time Reduced**: Thorough research on relocation could take twenty to twenty-five hours over several weeks as you read articles, watch videos, seek advice on forums, and talk with friends. Most answers take a chatbot a few seconds; within ten to fifteen minutes, you'll have the most critical information and data points.
    - **Savings:** *Over twenty hours* of research and conversations.

Energy and Stress

- A decision such as relocating your career and life will never come easy. However, quick access to an intelligent, insightful, and never-unavailable AI advisor can reduce the strain, especially when it's extended over time and conversations.

Tools and Tips

- The leading chatbots—**ChatGPT**, **Gemini**, **Claude**, **Llama,** and **Perplexity**—are all strong with general and specific queries such as relocation advice.

- There are also several chatbots designed for more emotional support, such as **Replika**, **You**, and **Woebot.**[35]
    - These chatbots can provide companionship, boost your mental health, and give answers to countless questions.

The Knowledge Revolution

AI chatbots are transforming the landscape of information retrieval and decision-making. By providing diverse, instantaneous insights, they help you confidently make informed decisions. They accelerate the process, provide substantial financial benefits, conserve emotional energy, and reduce opportunity costs. As AI continues to evolve, the potential for chatbots to serve as personal advisors in every aspect of life only grows, making them indispensable tools in our lives.

---

[35] https://www.replika.ai; https://www.you.com; https://woebothealth.com

CHAPTER 15

A CLOSER LOOK AT AI CHATBOTS

We discussed large language models (LLMs) at the beginning of the book and have seen their prowess throughout. These powerhouse AI platforms are exposed to us through the beautiful simplicity of a text input box, with the option of attaching documents, images, or audio files. This makes them as accessible and easy to use as any technology ever made. Yet, the power beneath them is incredibly vast.

These sophisticated AI systems can understand, generate, and interact with human language, offering invaluable assistance across various domains. In this chapter, we examine the nuances of the different LLM chatbots, guiding you through their unique features, helping you make informed decisions about which to use, and exploring the myriad of tasks they can accomplish.

LLMs: A Comparative Analysis

1. ChatGPT (OpenAI)

How it Works:

- Architecture: Transformer-based model—GPT stands for "Generative Pre-Trained Transformer."
- Training Data: A large dataset of internet text, fine-tuned for conversational capabilities.
- Fine-Tuning: Can be fine-tuned for specific tasks and domains.

Unique Aspects:

- Explicitly designed for generating human-like dialogue.
- Easily accessible through Open AI's Application Programming Interface (API), facilitating integration into various applications.
- Good at maintaining context over multiple turns in a conversation.

Specialty Tasks:

- Customer Support: Excellent for generating responses in chatbots and virtual assistants.
- Content Generation: Versatile in producing coherent and contextually relevant text.
- Creative Writing: Capable of creating engaging and innovative content.
- Educational Tools: Useful for tutoring and interactive learning scenarios.

2. Gemini (Google DeepMind)

How it Works:

- Architecture: Based on Google's Pathways Language Model (PaLM), extended with multimodal capabilities.
- Training Data: Uses diverse datasets, including text and images, leveraging Google's Pathways system to extract information.
- Multimodal Capabilities: Can process and understand both visual and textual data.

Unique Aspects:

- Combines vision and language understanding, enabling tasks requiring text and image analysis.
- Improved ability to interpret and generate responses based on complex, multimodal inputs.
- Seamless integration with Google's ecosystem, including Google Cloud.

Specialty Tasks:

- Ideal for applications requiring text and image inputs, such as generating image descriptions.
- Suitable for a wide range of applications, including enterprise use.
- Effective in interactive learning environments that use multimedia.
- Strong performance in tasks that require detailed analysis and comprehension of mixed data types.

3. LLaMA (Meta AI)

How it Works:

- Architecture: Transformer-based model designed to be efficient and scalable.
- Training Data: Trained on a large and diverse dataset focusing on efficiency.
- Performance: Aims for state-of-the-art performance in language understanding and generation.

Unique Aspects:

- Designed to perform well with fewer resources compared to other large models.
- Balances transparency with performance, promoting research accessibility.
- Suitable for large-scale applications and enterprise use.

Specialty Tasks:

- Ideal for academic and research purposes due to its efficiency and openness.
- Effective for large-scale deployments where resource efficiency is crucial.
- Strong performance in generating coherent and contextually relevant text.
- Capable of handling detailed and nuanced customer interactions.

4. Perplexity AI

How it Works:

- Architecture: Utilizes a combination of transformer-based models and search capabilities.
- Training Data: Leverages internet text and search data to provide answers and generate text.
- Integration: Combines language model capabilities with information retrieval techniques.

Unique Aspects:

- Uses search data to enhance the accuracy and relevance of responses.
- Provides up-to-date information by integrating search results into its responses.
- Designed to deliver precise answers to user queries by leveraging external information sources.

Specialty Tasks:

- Excellent for tasks requiring up-to-date and accurate information.
- Effective in providing detailed and relevant responses to user inquiries.
- Capable of generating text that is informed by the latest data and trends.
- Useful for detailed analysis and fact-checking tasks.

5. Mistral AI

How it Works:

- Architecture: Advanced transformer-based architecture focusing on high performance.
- Training Data: Trained on a large and diverse dataset to cover various domains.
- Specialization: Designed for high accuracy and performance in specific tasks.

Unique Aspects:

- Optimized for high accuracy and efficiency in language tasks.
- Can be fine-tuned for specific domains, enhancing its relevance and precision.
- Suitable for both small-scale and large-scale applications.

Specialty Tasks:

- Excellent for tasks requiring specialized knowledge and precision.
- High performance in generating detailed and contextually accurate text.
- Capable of handling complex queries with high accuracy.
- Strong performance in tasks requiring detailed and precise analysis.

6. Bing Co-Pilot (Microsoft)

How it Works:

- Architecture: Utilizes transformer-based models integrated with Bing search capabilities.
- Training Data: Combines web search data with language model training data.
- Integration: Deep integration with Microsoft's products and services, including Office and Azure.

Unique Aspects:

- Enhances responses with real-time search data from Bing.
- Seamless integration with Microsoft's suite of tools and services.
- Provides up-to-date information by leveraging Bing search results.

Specialty Tasks:

- Ideal for tasks requiring current and accurate information.
- Effective for integration into business workflows and productivity tools.
- Capable of delivering detailed and relevant responses by integrating search results.
- Can generate content that is informed by the latest web data.

7. Claude (Anthropic)

How it Works:

- Architecture: Transformer-based model focusing on safety and reliability.
- Training Data: Trained on a diverse dataset emphasizing ethical considerations.
- Safety Measures: Incorporates robust safety features to mitigate harmful outputs.

Unique Aspects:

- Prioritizes safety, transparency, and ethical considerations in its design and operation.
- They are designed to reduce the risk of generating harmful or biased content.
- Emphasizes research accessibility and transparency in its development.

Specialty Tasks:

- Ideal for applications where safety and ethical considerations are paramount.
- Provides reliable and safe responses, minimizing the risk of harmful outputs.
- Capable of generating high-quality text with a focus on ethical standards.
- Useful in environments that require safe and reliable AI interactions.

One common feature across these LLMs is "Transformer Architecture." Introduced by Vaswani et al. in their 2017 paper "Attention is All You Need," this neural network design has become the foundation for most advanced language models, including GPT.[36] It has led to rapid breakthroughs in the last few years—**revolutionizing** is not an exaggeration.

Beyond the Questions and Answers

LLM chatbots can perform an array of tasks, not just answer questions. These include:

1. **Content Creation**
    - They can generate original, high-quality content from publish-ready articles to programmatic code.
2. **Customer Service**
    - They provide real-time, 24/7 support, answering queries and solving problems.
3. **Language Translation**
    - The chatbots will translate text across multiple languages with high accuracy.
4. **Education and Learning**
    - They can assist in learning new languages, tutoring in various subjects, and more.

[36] Vaswani et al. "Attention is All You Need." *Advances in Neural Information Processing Systems,* 2017, https://proceedings.neurips.cc/paper/2017/file/3f5ee243547dee91fbd053c1c4a845aa-Paper.pdf. Accessed 23 Aug. 2024.

Savings

Subscribing to a single chatbot can offer significant savings across many aspects of personal and professional life. A single $10 to $20 monthly subscription can handle many of the financial, time, and additional benefits to physical and mental wellness detailed in this book.

In many cases, where a recommendation to a specialized AI app is made, the chatbot alone can provide sufficient insight or content. As a result, they can deliver a good portion of the *thousands of dollars* in annual financial savings and *weeks* in time savings, which add up across the categories covered in this book.

If you invest in a subscription to get the full benefits of these incredible tools, one primary consideration is image production. Most free tiers won't provide this feature or will be very limited. The ability to produce visuals is a powerful and magical tool that enables or enhances many categories.

Tools and Tips

Here is where you can find each of the chatbots.

- **ChatGPT** @ ChatGPT.com
- **Gemini** @ Gemini.Google.com
- **LLaMA** @ LLaMA.Meta.com
- **Perplexity** @ Perplexity.ai
- **Mistral** @ Mistral.ai
- **Bing Copilot** @ Bing.com/Chat
- **Claude** @ Claude.ai

Choosing a chatbot can be difficult for those not savvy about the nuances of each LLM. Fortunately, some great tools aggregate responses of multiple LLMs into one window. Not only do you get the instant response from multiple "genius" AI brains, but watching six streaming outputs at once is quite mesmerizing.

- **Chathub** we mentioned in a previous chapter. It is a simple Chrome extension that provides two windows for free when comparing LLMs at a time and six windows simultaneously for a small lifetime fee.
- **Playground AI** has various LLMs, including ChatGPT, Stable Diffusion (for image creation), and others specializing in certain types of content generation.[37]

Limitations and Validation

While LLM chatbots offer remarkable capabilities, they have limitations. They might need help understanding context perfectly, generate incorrect or biased information, and lack emotional intelligence. Occasionally, they make up information—or "hallucinate"—to satisfy a query. It's crucial to validate their output, especially for critical tasks, and combine their insights with human judgment.

If you are skeptical about something it says, ask them to provide the sources or confirm if what they said was fact or speculation.

[37] https://playgroundai.com

Leveling Up Our Game

LLM chatbots are transforming how we interact with technology, offering personalized, efficient solutions across a broad spectrum of tasks. By understanding the strengths and limitations of each model, users can choose the most appropriate chatbot for their needs, unlocking new possibilities for saving time and money while navigating the digital world more effectively.

SECTION V

ANYONE CAN CREATE

CHAPTER 16
PERSONALIZED GREETINGS

The holiday season is a time of warmth, connection, and sharing joy with friends and family. However, it can also be a period of significant stress, especially when sending out personalized holiday cards. Crafting a unique message for each recipient, managing a lengthy mailing list, and ensuring everything is sent out on time can be a formidable task.

AI can transform the traditionally labored process of creating holiday cards into an efficient, enjoyable, and personalized experience. Let's see how.

Add Joy to the Holiday Season

We all want to make holiday cards feel special, reflecting sincere warmth and wishes for a new year. We love getting personalized messages, but it's daunting to do it ourselves.

The idea of AI doing the writing for you may seem like a turnoff—can it really be personal? Yes, it can.

We need to see how AI facilitates and enhances what we do, not replaces it. We can keep our voice and our genuine intention while being more efficient. If, like many of us, you find yourself guiltily doing nothing more than signing card after card, then leaning on AI for assistance can make the effort of card sharing a deeper and more meaningful experience.

Tooling for Special Greetings

The first time set up for personalized holiday greetings will take a little extra effort, but far less than trying to write twenty-five or more personalized messages every year. In the coming chapters, we'll even show you how you can turn custom greeting cards into a business if you choose!

Select the App

There are several platforms where you can create your card, but **Canva** stands out for its integration of AI and numerous features in its Pro plan that make it worth the expense. You can get a free trial for thirty days and then pay monthly. So, the cost will be limited if you do it once a year. Some other apps are listed in the **Tools and Tips** section if you want to explore them.

Step-by-Step: From Design to Delivery

AI Tool: Canva

1. Create Your Card Design

A. Go to Canva and create or log in to your account.[38]

B. Choose a greeting card template or create a new design.

- There are thousands of designs which can all be custom-edited.
- Search templates for any theme, holiday, or event.
- All messaging, format, style, and imagery can be modified in any design.

C. AI features allow you to prompt imagery, ask for a recommended copy, or offer different design variations.

D. Add Text Fields that include "Name" and "Message". Place and format these text boxes on the card.

2. Prepare Your CSV File

A. Open a spreadsheet program such as Excel, Google Sheets, or any spreadsheet software.

Hint: Exporting your contacts into a spreadsheet can save you time adding all the names. Any contacts program will allow such an export.

38 https://www.canva.com/

B. Create your columns by labeling the first row with the fields you want to personalize. These should include:
   i. Full Name
   ii. To
   iii. Address
   iv. Description
   v. Message

If you exported contacts, then be sure to label the columns with the appropriate heading and delete the columns you don't need.

C. Enter your data by filling in the rows with the proper information.
   i. If you want AI to help create your message, use the Description box to provide some basic information. This can include the person's relationship with you and other family members or any other details that will allow the AI to create a personal message. Mention the milestones they hit that year, the obstacles they overcame, or simply the names of their kids and pets.
   ii. Don't worry about formatting these notes. They can be words, phrases, sentences, keywords. The revolution in AI is about understanding human natural language, and it can make sense of whatever words you put in the description box.

D. Save the file in CSV format (e.g., personalized_cards.csv).

3. Let AI Create Personal Greetings

A. Using a chatbot such as ChatGPT, attach the CSV file to your prompt and tell the AI to create personalized messages for the occasion based on the Description box.

- While most chatbots will understand the column headings, adding the column letter in the prompt is often helpful.
- For example, the prompt would be: "Make a personal Happy Holidays message in Column E 'Greeting' addressed to the person in Column B 'To' using the information in Column D 'Description.' Make a new CSV in the same format with the personal greeting added to Column E 'Greeting.'"
- If you want the message addressed to multiple people, simply include multiple names in the "To" column. For example, if it's a family of four, it might be "Becca, Harold, Lizzy, Ben."

B. Submit the prompt, and let the chatbot produce a new CSV with personalized greetings.

C. Review the CSV and edit the greetings as needed. You can also edit the descriptions field and ask the chatbot to take another pass at the messages.

4. Use Bulk Create to Personalize Cards

A. Return to Canva and access Bulk Create:

- Click on "Apps" from the left sidebar in Canva.
- Search for and select "Bulk Create."

B. Upload your CSV File:
   - Click on "Upload CSV."
   - Select your CSV file (personalized_cards.csv) and upload it.

C. Map Data Fields:
   - Map the columns from your CSV file to the corresponding text boxes in your design.
   - Be sure to include the "To" and "Greeting" fields.
   - Canva will automatically generate multiple versions of your design with the personalized data.

5. Review and Adjust Designs

A. Review the generated designs to ensure the personalization of each card is correct.

B. Edit any individual card if needed for better alignment or formatting.

C. If you want to put in a little extra effort, now is the time to make even more personalized edits to the card. AI recommendations in Canva can quickly modify the text and any image or style.

6. Print Your Cards

A. Choose Print Option:
   - Click on the "Print Cards" button.

B. Select Printing Options:
   - Choose the type of card, paper quality, finish (matte, glossy), and quantity.

C. Review Your Order:
   - Review the final designs and print options.

D. Place Your Order:
   - Enter your shipping and payment details.
   - Confirm and place your order.

7. Delivery

A. Shipping: Canva will print and ship the cards to your specified address.

B. Receive Your Cards: Wait for the delivery, and your personalized greeting cards will arrive ready for distribution.

8. Address and Send

A. All you need to do now is address the envelopes. If you include addresses in your CSV, this becomes a valuable data source.

A Forever Template

As mentioned, setting up a CSV the first time will take a little work, especially providing descriptions of the people you are sending the greeting to. However, this now becomes an excellent template for the future. Each year, you can simply tweak any of the descriptions, maybe including something relevant to the year just past, and get a whole new set of personal messages.

Adding new names to the list and creating a new greeting card with updated photo(s) is also very simple.

For the coming decades, you will be known for always sending personal notes during holidays, bringing greater joy to the holidays, and making them more meaningful for those in your circle of family and friends.

Power of AI in Personalization

Combining design tools and chatbots makes custom greeting cards fast, easy, and creative. Here are all the things AI can do to support your creative efforts:

1. **Template Suggestions**
   - Analyze your preferences and suggest design templates that match your style and occasion.
   - Modify existing templates based on input, ensuring the design fits your needs.
2. **Personalized Messages**
   - Generate personalized messages based on the recipient's name, relationship, and occasion.
   - Analyze the sentiment you want to convey and suggest appropriate wording.
   - Adjust the tone, length, and style of the message to fit the occasion, whether formal, casual, humorous, or sentimental.
3. **Design Assistance**
   - Suggest optimal layouts for your card, balancing text and images for visual appeal.
   - Recommend color schemes that complement each other and match the occasion.
   - Suggest font combinations that look good together and fit the card's theme.

4. **Image Integration**
    - Select and suggest the best photos from your collection to include in the card.
    - Automatically remove backgrounds from images, making incorporating photos into the design easy.
    - Enhance the quality of images, adjust lighting, and apply filters to match the card's aesthetic.
5. **Bulk Personalization**
    - Help personalize multiple cards simultaneously using data from a spreadsheet.
    - Create unique cards for each recipient efficiently.
6. **Design Automation**
    - Provide intuitive drag-and-drop interfaces, making it easy to customize card elements.
    - Automatically resize designs to fit different card sizes or formats, ensuring consistency across mediums.
7. **Creative Elements**
    - Generate or suggest unique artwork, illustrations, and graphics that fit your theme.
    - For digital cards, add animations to make the card more interactive and engaging.
8. **Real-time Collaboration**
    - Platforms like Canva allow multiple users to collaborate in real-time, making it easy to get feedback and adjust quickly.

    - Manage different versions of your design, allowing you to track changes and revert to previous versions if needed.

9. **Printing and Distribution**
    - Optimize your design for printing, ensuring high-quality output.
    - Allow direct ordering from within the design tool, streamlining the printing and delivery process.
    - Help with address management, printing envelopes, and even sending cards directly to recipients.

Savings

Time

The real value of AI in creating personal greeting cards is in time. If you like to include personal messages and designs, it can save you several hours, if not weeks.

- **Personalized Messages:** Writing a personalized paragraph could take ten to fifteen minutes. Describing quick summary points might take two minutes, with the AI generation of messages completing the message in seconds.
    - **Savings:** five to ten if you write twenty-five to fifty holiday cards, as the time for each is reduced to a couple of minutes of notes.

- **Creation:** Using an existing template can save hours in design. If you enjoy design, the speed at which you can go through creative options enables you to explore more options much faster.
    - **Savings:** The ultimate creative choices are yours; AI simply empowers you to expand your vision and options. *Hours* will be saved either way.
- **Organization and Distribution:** Managing large mailing lists and card production can be tiring and prone to errors. Combining digital tools and AI analysis can ensure fewer problems and faster resolution.
    - **Savings:** one to two *hours* for a set of cards.

Estimated Net Time Saved: It depends on the quantity you produce and the creative effort you want to make, but AI can cut your workload to one-tenth of the usual time.

Money

- **Personal versus Generic:** Let's be honest. AI tools and printing services will likely cost a bit more than buying cards and writing personalized greetings by hand. However, the level of personalization and number of people you can reach will be far greater—and save your writing hand much grief!
- **Low Total Costs:** The costs will vary depending on your chosen services and quantities. However, it's likely that you can create, personalize, and send greetings to all the important people in your life for less than $100.

While the cost to purchase, sign, and send a set of pre-made cards may be less, the value of the personalization at scale is immeasurable and well worth it.

Stress

- **Anxiety Savings:** The cost to our emotional well-being, especially during the holidays, should not be ignored. Holidays bring a lot of responsibilities, and utilizing tools that can reduce the burden while enhancing meaningful connections is a powerful addition to our lives.

Tools and Tips

- **Canva** is excellent for designing, batch messages, and printing.[39]
- **ChatGPT** can populate a spreadsheet with personalized messages.[40]
- **Vistaprint** is excellent for printing and mailing services. Vistacreate allows for design with AI, although personalized messages must be copied and pasted into each and not uploaded in batches.[41]
- **Appypie** heavily relies on AI to create personalized cards. It is suitable for digital-only cards. Personalized messages will need to be inserted in, not batch uploaded.[42]

---

39 https://www.canva.com/

40 https://chatgpt.com/

41 http://www.vistaprint.com

42 https://www.appypie.com/design/

Improving the Quality of Our Connections

By leveraging AI tools, anyone can create custom, heartfelt greetings for any occasion, making the recipients feel valued and loved. We should not be ashamed of failing to find the time or energy to send personalized greetings. We all lead busy lives with work and family commitments, which generally become more intensive during the holiday season or around special events. Turning to AI tools to enable us to do more of what we truly care about with less strain can unlock more value and connection in life.

CHAPTER 17

LAUNCH A WEBSITE

The tools to make a website have advanced considerably over the last decade. What once required skilled designers and could take months is now available in minutes. This is primarily thanks to AI, and almost nowhere do we see its speed and power in action better.

Why do you want a website, you might ask? This may not be a question for some of you, but there are countless possibilities for those who have never had one or even considered it. Rather than list them, I suggest you use one of the sites recommended in this chapter to explore the possibility. With as few as a couple of prompts, AI-powered website builders provide you with a site design for anything you can imagine. I'd wager that once you experience how easy it can be to create attractive pages, express ideas, generate images, claim your domain, and sell almost anything, you'll find yourself inspired to explore deeper.

In the next chapter, we'll look at tools to generate blog or article content quickly and regularly, so if that is the type of website you'd love to publish but are intimidated by how to write the content, AI has a solution for you there, too!

Choosing a Tool

As with every place AI is present, there are numerous options and more all the time. Here are three options to consider, but keep your eyes open for more. As with everything, your AI chatbot of choice is one of the best sources to query for the latest and greatest.

1. **Wix** has been in the website-building game for a long time and has been at the forefront of integrating AI. It asks you a few more questions and seeks a bit deeper understanding of your thinking than some other new tools, but it offers a robust site vision for the extra couple of minutes.
2. **Launched** simplifies the process even further, asking just a couple of questions before rendering a template that can be surprising and inspiring.
3. **Mixo** is another rapid creation tool. With just an idea and a color choice, it will render a site template in seconds.

All of these allow creating templates for free, so dive in and try it. They also enable quick regeneration of the template,

Step-by-Step: Website Creation

For the purposes of this explanation, let's imagine we were so inspired by creating personalized greeting cards that we decided we should turn it into a business. What do you need for that? A website. Let's see how, within minutes, we can have our business up and running.

AI Tool: Wix

1. Design It

Follow the prompts of the tool you chose, and within minutes, if not seconds, you will have a site template. This prompt could be as simple as: "A destination to sell personalized greeting cards."

It's always good to add a little extra detail to your prompt, such as: "The website will be a place for me to sell my custom-made greeting cards and allow people to order personalized greeting card designs that I'll fulfill. The site should feel personal, warm, and artistic. It should inspire people to want to send personalized greeting cards."

You can also prompt it to add components, such as "Add a Blog page where I can make suggestions on how to create your own cards" or "Include a community section where people can provide what kinds of cards they want to be made."

2. Choose a Domain Name

Website creators have integrated domain acquisition into their workflow. They usually include a domain in a base subscription of about $10 per month. You can do this immediately, or if you're not ready, skip it and continue preparing your site.

When you're ready to choose a domain name, make it memorable, relevant, and reflective of your mission of spreading cheer. Consider using keywords in your domain like "greeting," "happiness," or "holiday" to resonate with your cause. The domain checker will quickly check its availability and may offer alternatives if your specific request is not available.

Tip: When it comes to domains, a ".com" address may be difficult to find or costly, but there are numerous other TLDs (Top-Level Domains) available, such as ".net," ".co," ".biz," ".shop" etc. which can be more available with the precise name you want, and much less expensive.

3. Customize Your Design

The website builders will render the template of a complete site, including a shop, a blog, imagery, and even business names. If you don't like it at all, have it regenerate a new one.

- Every part of the template can be customized. Click a section, and tools will pop up to edit text, change images, adjust color palettes, modify text styles, and more.
- Navigate the different parts of the site and customize each section. If you don't want a section or page, delete it.
- Add Features such as forms for visitors to request greeting cards and provide recipient

details. Wix, for example, offers applications like Wix Forms for this purpose.

- Not sure what you want in any given section? Ask the AI to change just that piece.
- Not feeling creative or not your thing to delve into specific design? Keep it as is. You've spent a couple of minutes and have a professional-looking site.

4. Add Content

Seeing and exploring your template could provide much inspiration for what you want to brand and include on the site. However, this part can also become intimidating. That's when you turn to AI for help.

- Utilize AI-powered tools like ChatGPT and other chatbots to generate engaging content. Prompt it to make welcome messages, About pages, or greeting card descriptions.
- Remember, with the AI chatbots, prompt them in natural language to provide the best and most customized result.
    - For example, "Write a welcome message for my website that will inspire people to want to order personalized greeting cards as a way to make more meaningful connections with their friends and families over the holidays."
- Use your chatbot to help develop a brand name for your company and site.

- In the next chapter, we'll discuss including blog posts or articles, which can be a great way to engage people on your site and market a business. This, too, will turn an intimidating task into a very doable one, turning hours of weekly writing into minutes of AI creation and light edits.

5. Swap Images

Your site will include imagery. You can change any of the images by selecting them and requesting that a new one be generated.

- Some site builders will use AI generation, which will require a subscription at this step (or once you use up the free credits).
- Other site builders have a library of images you can search with keywords.
- With every image, you can also upload your own, which will auto-format to the required dimensions.
- If you like them as provided, keep them!

6. Set up the Shop and Start Selling

One of the most exciting aspects of website builders is that they have built-in e-commerce. You will need to subscribe to set it up, but once you've done that, it's only a few clicks to set pricing, swap the placeholder images for your greeting cards, and allow people to purchase!

You can begin with as simple as a few card templates you've designed that are blank inside. You can also sell products such as "Custom Orders," where people can request a design and messaging. You can further offer the service of education by telling them what you learned in the last chapter on how to personalize messaging for a list of family and friends.

7. Review, Edit, Optimize

Once you've read through and completed any final edits, you can preview the site on desktop and mobile layouts. The website builders are generally very good with adaptive design for different devices, but you may decide to make a few more tweaks to the design so it's as mobile-friendly as possible. Mobile is where people increasingly use the internet to find and order products.

8. Connect Your Domain

If you haven't already chosen and secured a domain, this would be a good time to do it. Although you can technically launch your site with your website builder's domain, this would not be considered professional and would turn away most people. You also won't get the benefits of branding and SEO (discussed next).

As noted, you can often claim a domain inside the website builders themselves as part of the subscription, which will run around $10 per month for the initial starter level. Alternative ways to secure a domain are in the **Tools and Tips**.

9. Search Engine Optimization (SEO)

A crucial part of launching a website is making it findable by search engines. SEO is how you can improve the search engine's ability to discover your site. There is an extensive science (and some would claim art) to this, but as a starter, it begins with keywords embedded in your website header.

The website builders use AI to propose keywords based on your designed site and its purpose. You can also use chatbots to offer keywords. These are easily inserted into the text fields the website builders provide. Some website builders may also offer guides or tips on improving your SEO.

10. Publish Your Website

Now push a button, and you are live to the world. Even assuming you've spent some time customizing images and copy, you have likely turned your hobby into a business in less than one hour!

The Full Picture: Websites x AI

There are many things AI empowerment offers for website creation, including:

1. **Design Assistance**
    - Create websites based on user preferences and content input.
    - Suggest design improvements, color schemes, and layouts.
    - Enhance, resize, and optimize images.

2. **Content Creation**
   - Generate website copy, blog posts, product descriptions, and other content.
   - Personalize content for individual users based on their behavior and preferences, enhancing user engagement.
3. **Coding and Development**
   - For those who want to go beyond the templates into more profound web creation, AI will write code snippets, automate repetitive coding tasks, and suggest code completions.
   - Detect and suggest fixes for coding errors and vulnerabilities.
4. **SEO Optimization**
   - Suggest keywords to improve search engine rankings.
   - Recommend improvements in structure, readability, and keyword usage.
5. **User Experience (UX) Enhancement**
   - Provide customer support, answer queries, and guide users through the website.
   - Analyze user behavior to improve navigation, layout, and content placement.
6. **E-commerce Integration**
   - Suggest products to users based on their browsing and purchasing history.
   - Manage stock levels, predict demand, and optimize pricing strategies.

7. **Performance Optimization**
    - Improve website loading times by compressing images, minifying code, and caching content.
    - Automate A/B testing to determine the most effective design and content variations.
8. **Security Enhancements**
    - Monitor and detect real-time security threats, such as malware and hacking attempts.
    - Identify and prevent fraudulent activities, particularly in e-commerce transactions.
9. **Analytics and Insights**
    - Provide insights into website performance, user demographics, and behavior patterns.
    - Predict future trends and user actions to help in strategic planning and decision-making.

Savings

Money

- **Web Design**: Hiring a professional designer can cost anywhere from $500 to over $5,000 for a custom website, even more if you get into advanced development. AI-driven platforms have free tiers and premium plans ranging from $10 to $39 per month, significantly lowering upfront costs.
    - **Savings:** *Hundreds to thousands of dollars.*

- **Content Creation**:

  Professional copywriting services can cost hundreds to thousands of dollars, depending on the volume and complexity of the content.

  - **Savings:** Even at $10 or $20 monthly for their premium plans, AI chatbots significantly reduce or eliminate these costs.

- **SEO Services**: Professional SEO services cost from $500 to over $2,000 per month for ongoing optimization.

  - **Savings:** AI-driven tools embedded in website builders are often part of the subscription cost, offering substantial savings.

- **Maintenance and Updates**: Professional website maintenance and updates can cost $50 to $100 per hour.

  - **Savings:** AI-driven platforms include these as part of their service, providing automatic updates and monitoring tools at no additional cost.

Estimated Net Money Savings: *Thousands of dollars* on the design, development, content creation, operation and maintenance of your site.

Time

- **Design**: Depending on complexity, a custom website could take weeks to months.

  - With AI-powered tools, a basic but fully functional website can be designed in *under an hour.*

- **Writing:** Content for your website, including card descriptions, about pages, and blog posts, could traditionally take several days to weeks.
  - With AI, you draft it *in minutes.*
- **Search Engine Optimization**: This is crucial for ensuring your website is discoverable. Manual SEO can be a slow and ongoing process, taking days to research and implement initial optimizations.
  - AI tools provide personalized SEO plans *in minutes.*
- **Performance Analysis:** Gathering feedback and adjusting is faster with AI analytics tools that provide real-time user engagement data. Manual analysis could take days or weeks.
  - AI tools can offer insights *in hours or even minutes.*

Net Time Savings: Websites involve many elements, from building to optimizations. When these are all combined, the savings can be *two to six months.*

Realistically, the savings in time and money for building websites is the difference between doing something and not doing it. Whether you want to share ideas with the world, promote your business, offer services, or sell your products, AI has unlocked an entirely new powerful resource for your life and career.

Tools and Tips

- **Wix, Launched,** and **Mixo** all have impressive AI tools for rapid creation and iteration. There are many more![43]
- **Squarespace** is likely a website builder you have heard about. It integrates AI deeply and is great for people with a more precise or detailed idea of what they want for the site.[44]
- If you want to get your hands dirtier with the design, the same **Canva** subscriptions you used to make personalized greeting cards can help you quickly design beautiful web pages.
- Any leading AI chatbots we've mentioned previously can provide written content. **Chapter 15** highlights which LLMs are optimal for it.
- While many website builders offer domain purchasing services, **GoDaddy** and **NameCheap** are popular choices that offer competitive prices and easy integration with various platforms. You can quickly secure your domain with them and attach it to your site through the website builder.[45]

Turning Your Hobby into a Business

If you followed this chapter, even purely as an exercise of curiosity, you have discovered how to create and launch a heartwarming greeting card website in less than an hour.

---

[43] http://www.wix.com; https://launched.site; https://app.mixo.io/

[44] https://www.squarespace.com/

[45] https://www.godaddy.com; https://www.namecheap.com/

Such a website serves as a platform for spreading holiday cheer and demonstrates AI's power to simplify the web development process.

For individuals and organizations looking to build a professional site without substantial initial investments, this is not just a choice but a transformative solution. Whatever your idea, and whether you're a beginner or have some experience, these tools will help bring your vision to life and enable you to impact your life, if not the world.

CHAPTER 18

BLOGGING AND PUBLISHING

In the fast-paced world of digital media, maintaining a daily blog and ensuring its distribution across various platforms can be overwhelming. With the advent of AI, content creators now have powerful tools to streamline this process. AI can assist in generating ideas, writing content, optimizing for search engines, and even publishing across social media and websites. This chapter explores how AI can revolutionize daily blogging by making it more efficient and far-reaching.

Become a Content Creator

You've been inspired to make your own cards with special messages, build a website to sell your designs, and offer the services of making personalized cards for others. Let's imagine that you want to take it a step further.

It could be to market your greeting card business, build your name, and draw people in with your content. Content publishing is a powerful tool for driving awareness and engagement with your brand.

In addition, blogging and other social media content could fit as part of your mission to show people the importance of personalized card giving on holidays and special occasions. By demonstrating how it can be done and even providing a service to help people do it, you are spreading good vibes worldwide.

Generating Content Overview

Whatever theme or subject(s) you choose for your blog and messaging, the objective is to generate engaging content consistently and distribute it across platforms like Facebook, Twitter, LinkedIn, and, of course, your website. Here's a process you follow:

1. Generate Ideas
2. Create the Content
3. Create Support Imagery or Animation
4. Optimize the Content
5. Publish and Distribute

These five steps will turn you into a content creation machine, enabling you to build an identity and develop a following for your ideas and business. Let's see how AI helps you each step of the way, making authoring material easy, fun, and successful.

Step-by-Step: Create and Distribute

1. Ideation and Planning

Tool: Chat GPT, Claude, or any LLM chatbot

Process: In the context of our Greetings business, you could ask the chatbot:

Provide a list of fifty ideas for blog posts about how personalized greeting cards and messages are important to society.

- This is just one of many prompts you can use about the subject.
- You can prompt for lists on card creation, trends in greeting cards, history of greeting cards, best words and phrases to express joy on the holidays, etc.
- In less than a minute, you'll have hundreds of ideas that can last you years of blogging.
- Put all the ideas in a document and choose which one to use on any given day.
- Organize the ideas into weeks or months, providing structure to your posts.
- Ask the chatbot on any given day to review the list for something relevant to the current calendar day or event happening in the world or to prompt it for a new idea.

2. Writing Content

Tool: Use the same AI chatbots that generated the ideas to write the blogs, social media posts, and any other content you want to produce.

Process: Choose one of the ideas and prompt the chatbot to write your blog content. Be sure to use the target number of words.

- If you post regularly (multiple times a week), 500 to 750 words is a good length. Regular posting is recommended to build SEO and a following.
- If you wanted to post articles instead (or on occasion), these would be 2000 words or more.
- Longer articles can help SEO as well. Making these a topic that other sites might link to could be very helpful in building your brand. Articles could be posted to a separate page or section of your website, different from the shorter and more consistent blogs.
- If you want to explore an AI tool specializing in writing content marketing, Article Forge is worth a look. It will cost an additional subscription, but you can use the free trial to test if you like the content better.[46]

Once you have your primary content, you may also want to produce a shorter version for different platforms. For example, take the blog text, paste it into the AI chatbot, and prompt:

Use this blog post and write three possible messages of less than fifty words based on the content made for X (formerly Twitter)

or

[46] https://www.articleforge.com/

Use this blog post and write three possible messages of less than 250 words based on the content made for Facebook

or

Turn this blog post into a script for a video for YouTube

AI tools produce all these formats in seconds, providing ample content to build your content marketing strategy. You can then edit the content yourself, or if you want to be more ambitious, there are additional AI tools for improving it. See **Tools and Tips** for options.

You might wonder why we didn't simply ask the chatbot to write a blog from the very start. To produce the amount of content you will need to sustain engagement—an effort that involves (at minimum) multiple posts a week—you'll create far more exciting content with greater depth by starting at the idea layer. Had we prompted, "Write a blog post about how personalized greeting cards and messages are important to society," we would have had one blog, not fifty.

Even better, use Chathub and prompt multiple AI chatbots for ideas. You will get numerous lists for each idea prompt. Remember, you need an exponential amount of content to be successful on the internet.

3. Image Creation

Tool: You'll want an AI Image generator to support your effort, but you will likely already have access to one. It could be with the subscription of your website builder, such as Wix or Canva's Magic Visual Generator (already part of your Canva subscription), ChatGPT (if you have the paid subscription), and Bing Create (if you have a Microsoft Office subscription), all of which give you access to image creation. Adobe Firefly is a good option if you are looking for a stand-alone option.

Process: Create visuals for your blog post by providing descriptive prompts to any AI-powered tool mentioned above. Imagery makes your post more engaging to users and is more robust for SEO. An image can be used across platforms, helping build your brand identity.

Use the blog's theme, keywords, and text from the post itself to prompt the AI. There is no need to come up with anything new; you already have it! The shorter social media versions act as excellent prompts, as they highlight the most compelling content.

Visuals can take many forms: aesthetically pleasing, data or charts, or samples from your business. In the example of our greeting cards, many of the supporting images for posts will presumably be in this card format, which will be a regular reminder of your core offering to your audience.

You can also vary the style by including it in your prompt. Photographic, illustrative, comic-book, impressionist, and more are all fair game for AI Image generators. Prompt for any artistic style from history or inspired by any artist, and you will get a rendering of that likeness.

4. SEO Optimization

Tool: If you're ready to invest further in your business, AI tools that help you improve your search discovery can be a worthwhile investment.

Process: Paid SEO tools are one of the higher-cost monthly subscriptions you'll find in this book. Popular tools such as **SEMrush** or **Moz Pro** can run $100 or more per month. There are cheaper options, such as **Ubersuggest** or **Mangools**, but these, too, can run upwards of $50 per month. A core focus is on improving the keywords in your post. Input your draft blog and post into the tools. They will suggest keywords, readability improvements, and other adjustments to ensure your post is discoverable and ranks well on search engines.

Your website builder's premium may already include some SEO keyword analysis for your blogs as well as the site overall.

If you don't want to invest in a paid tool immediately, there are still ways to improve your SEO and help your business get off the ground. Here are some free steps you can take:

A. Identify keywords that align with your Greeting Card business. A chatbot can help you.

B. For each Blog Post, craft a compelling title that includes your primary keyword(s). Keep it under sixty characters.

C. Write a persuasive meta description (quick summary) of under 160 characters that includes your primary keyword(s).

D. Include keywords in webpage headers to improve readability and SEO.

E. Ensure keywords are incorporated throughout your post, including in the introduction, headings, and conclusion. Be careful, though; you don't want redundancy to harm the quality of the post.

F. More comprehensive content of 1,000 words or more often ranks better in search. However, this should be balanced to make it easily consumable. Including longer articles a couple of times per week can be a good compromise.
G. Write clearly, concisely, and engagingly. Use short paragraphs, bullet points, and subheadings. Consider tightening the writing with tools such as **Grammarly** or **Hemingway App**.
H. Include keywords in the file name and alt text of your images.

Even without a paid tool, remembering these and getting in the habit of including them with each post can improve your SEO. A free AI chatbot can help you do all the above.

Once you start making some money, accelerate your business with a paid tool. Some, like **Surfer AI**, will write keyword-optimized content for you and support SEO.

5. Publish

Tool: Nothing special should be necessary—simply add the text and image to your website!

Process: Post the text and images and save! Depending on the website builder, blogs may need an added page. Assuming you have a base subscription for your site, this should be at no additional cost.

Twitter and **Facebook** channels can be valuable to building your brand and business. Add your optimized word count versions for these platforms. If you make compelling visuals, such as beautiful greeting cards, **Instagram** can be a key promotional channel.

Posting content should take only a few minutes a day. Tools such as **Buffer** and **Hootsuite** allow you to schedule and automate posts. You could set up all your posts for a week in one day and let the tools do the work. These tools also provide analytics for how well people engage with your content.

There will be monthly costs, however, so it depends on how much investment you are prepared to make. While convenience is nice, it is best to prioritize investing in tools that can improve your content and SEO.

Beyond Content Creation: What AI Can Do

Capabilities extend to other areas that may come in handy as you grow your blog and business:

1. Automated content planning and calendar scheduling.
2. Sentiment analysis to gauge reader response and tailor content accordingly.
3. Performance analytics to understand what content performs best.
4. Automated responses to comments and messages to foster a community.

Savings

Money

- **Content Creation**: Hiring a professional writer can cost anywhere from ten to fifty cents per word, resulting in $100 to $500 for a 1000-word article. Using AI writing assistants can significantly reduce this cost.
    - **Savings**: *$100 to $500 per post.*

- **SEO Services**: SEO consultants might charge $75 to $150 per hour. Using SEO tools can provide similar insights at a fraction of the cost.
    - **Savings**: *Up to $300 per article* for SEO analysis and optimization.

- **Graphic Design**: Hiring a designer for custom visuals can cost between $50 to $200 per image. AI Tools allows you to create visuals at a much lower subscription cost.
    - **Savings**: *$50 to $200 per article.*

Net Money Savings: *$150 to $1000 per post*, depending on the depth of outsourcing. If you post multiple times a week, you will save *thousands of dollars monthly.*

For some of you, the costs for new subscriptions may be all new. However, if you are launching a new business, these startup costs are a fraction of what they would have been pre-AI content creation, enabling you to explore opportunities that were once well beyond reach.

Time

- **Ideation and Planning**: Traditional brainstorming sessions could take hours or even days. AI can reduce this to minutes.
  - **Savings**: *Up to three or four hours.*
- **Draft Writing**: Writing a draft from scratch could take four to eight hours, depending on the article's depth. AI writing tools like ChatGPT can produce a first draft in minutes, with a bit of added time for editing (the AI takes seconds, and the minutes come from manual revisions).
  - **Savings**: *three to seven hours.*
- **SEO Optimization**: Manually optimizing for SEO involves extensive keyword research and adjustments, which could take two to three hours. AI Tools provide instant recommendations.
  - **Savings**: *Two hours.*
- **Image Creation**: For someone with intermediate design skills, designing visuals could take two to three hours. An AI-powered tool reduces this to minutes.
  - **Savings**: *two hours*
- **Proofreading and Editing**: Manually proofreading and editing can take up to two hours. Grammarly or the Hemingway App can cut this down significantly.
  - **Savings**: *An hour and a half.*
- **Publishing and Promotion**: Scheduling social media posts and publishing content manually can be

time-consuming. Automation tools can do this in a fraction of the time.

- o **Savings:** *One hour.*

Net Time Savings: Approximately *ten to fifteen hours per blog post*, which adds up to *over one hundred hours monthly* for regular posting.

Tips and Tools

- **AI Writing Assistants:**

  Look for tools like **ChatGPT**, **Claude**, and other AI chatbots for content creation.

  Specialized tools such as **Autoblogging AI**, **ArticleForge**, and **SurferSEO** cost a bit more but deliver SEO optimization as well and should be considered if you want to grow your audience.[47]

- **AI Editing:**

  Grammarly @ app.Grammarly.com or **Hemingway App** @ HemingwayApp.com offer quick AI-based edits to optimize readability and overall writing quality.

- **AI SEO Tools:**

  SemRush and **Moz Pro** are more costly, **Ubersuggest** and **Mangools** somewhat less, but all are significant monthly subscriptions.[48]

---

47 https://autoblogging.ai; https://www.articleforge.com; http://www.surferseo.com

48 http://www.semrush.com; https://moz.com;

- **Social Media Management:**

  Platforms like **Buffer** or **Hootsuite** offer AI capabilities for scheduling and analytics.[49]

- **AI Image Generators:**

  If you subscribe to an AI chatbot, check there first. **ChatGPT**, **Gemini**, *and* **Bing** provide AI image creation as part of their premium plans. **Canva's** *Magic Visual Generator* comes with that subscription. **Adobe Firefly** is worth a look if you are looking for a new tool from scratch.[50]

- **AI Animation**:

  If you really want to make a splash, nothing beats animation on your website or social posts. Text-to-Animation, or Image-to-Animation, is becoming increasingly available. Both OpenAI with **Sora** and Google with **Veo** are preparing such tools. **Runway AI** already enables short text-to-animation at little or no cost.[51]

- **AI Avatar:**

  Now we're really leveling up. Use your AI chatbot to make a script for your blog, and then jump over to **HeyGen**, where you can design a human avatar to deliver your script. Post this on YouTube or TikTok to expand

---

https://neilpatel.com/ubersuggest; https://mangools.com

[49] https://buffer.com; https://hootsuite.com

[50] http://www.adobe.com

[51] https://www.soraapp.com; https://deepmind.google/technologies/veo; https://runwayml.com

your reach. It can even speak multiple languages, turning your little website into a global business.[52]

As you can see, those few minutes spinning up a website have opened the door to someday running a worldwide, multi-language global enterprise. Even if you keep it as a hobby or simply a platform to express ideas and connect with others, AI can unlock numerous ways to merge your style, voice, and ideas with rapid content creation.

Uplifting Your Voice

Incorporating AI into daily blogging and social media distribution workflows can significantly enhance efficiency, creativity, and reach. By leveraging AI for idea generation, content creation, SEO optimization, and distribution, bloggers can maintain a consistent presence across multiple platforms with minimal effort.

This chapter has shown that AI is not just a futuristic concept but a practical tool for content creators seeking to optimize their digital presence and engage with an audience daily. And that content creator can now be you!

52 https://www.heygen.com/

CHAPTER 19
CRAFTING COMICS

Comic creation is an art form that combines storytelling with visual artistry. It's a medium that has entertained and informed for centuries, from classic newspaper strips to modern webcomics. Today, AI technologies offer new possibilities for creators, making comic creation more accessible and efficient.

Comics can launch memorable characters and tell stories about anything. They can be funny, dramatic, shocking, and sad. Whether you're an aspiring artist, a seasoned creator looking to streamline your process, or have never really considered it but want to explore it as a hobby and form of expression, AI can be a collaborator in bringing your characters and stories to life.

Step-by-Step: From Concept to Comic Strip

1. Ideation and Character Creation

Imagine you have an idea for a comic but need help with the design and backstory of your main character. You turn to an AI-powered character generator like the free **Perchance, Artbreeder,** or **GetMerlin** character generators. Each allows you to describe the character, add features, and generate original portraits. Perchance has a long list of different styles to choose from, allowing you to work with any era, genre, or texture.

2. Writing Dialogue and Storylines

With your character(s) designed, crafting engaging storylines and dialogue is the next step. **Plot Generator** and **Toolbaz** are free tools to help generate ideas or refine dialogue. **Wordkraft AI** and **Squibler** have free tiers with more options for upgrading to premium. You input the basics of your story, and the AI suggests dialogue options, plot twists, or even entire narratives based on your input.

3. Drawing and Illustrating

AI drawing tools like **DALLE,** which comes with the ChatGPT premium tier, or **Canva** provide tools for comic strip imagery. **Fotor** and **AI Comic Factory** are also good tools to consider. For instance, you can describe a scene where your character is battling a villain in a futuristic city, and the AI will create a visual representation for you.

4. Assembling the Comic Strip

With panels created, it's time to put together your comic strip. AI software like **Comic Draw** or online platforms like **Canva** offer templates and tools to assemble your comic. You can adjust layouts, add speech bubbles, and edit visuals all within these platforms once AI does the initial design.

5. Publishing and Sharing

Finally, you're ready to share your comic with the world. Platforms like **Webtoon** or **Tapas** are great for reaching a broad audience. These sites cater specifically to comic creators and fans, offering a community where you can publish your work and receive feedback.

6. Daily Creation

The real power of AI in comic creation shines when you produce new content regularly. Using the same character and AI tools, you can quickly generate new scenes, dialogues, and story arcs. This process significantly reduces the time it takes to produce each comic strip, allowing for daily updates if desired.

The Story of Cybel: The Hybrid Hero

Let's delve into a specific example of creating a comic strip using AI featuring a character named Cybel. Cybel is a unique being, half-human and half-machine, grappling with the duality of their existence through internal debates between their human empathy and machine logic. This storyline offers rich narrative potential, exploring themes of identity, ethics, and technology integration with humanity.

1. **Character Creation**

   Using **Artbreeder**, we craft Cybel's appearance to visually represent their dual nature. The human side shows softer, organic features, while the mechanical side displays sleek, cybernetic enhancements. This visual contrast helps immediately convey the character's internal conflict to the audience.

2. **Writing Dialogue and Storylines**

   For our first comic strip, we decided on a scenario in which Cybel faces a moral dilemma: save a malfunctioning robot slated for decommissioning or respect the laws of the city, which sees it as mere property.

   We use **ShortlyAI t**o brainstorm dialogue that captures Cybel's internal debate. The AI suggests lines reflecting the logical, emotionless machine and empathetic human viewpoints. This combination of dialogue showcases Cybel's inner struggle.

3. **Drawing and Illustrating**

   To create the panels for our comic, we turn to **DALL·E**. We input detailed descriptions of each scene, like "Cybel standing at a crossroad, half their face illuminated, reflecting their human side, while the other half is in shadow, showing the mechanical features. A small, broken robot lies at their feet under a streetlamp." DALL·E generates visuals that capture these moments, providing unique artwork for our comic.

4. **Assembling the Comic Strip**

   Using **Canva**, we select a comic strip layout and upload the AI-generated artwork. We carefully place the dialogue into speech bubbles, ensuring the flow of conversation mirrors Cybel's internal conflict. The design allows us to visually juxtapose the human and machine aspects, enhancing the storytelling.

5. **Publishing and Sharing**

   Once our comic strip, "The Duality of Cybel," is complete, we publish it on **Webtoon**. The platform's large audience of comic enthusiasts provides immediate feedback, with many resonating with Cybel's dilemmas and the exploration of themes related to AI and humanity.

6. **Expanding Cybel's Universe**

   Using the established character and setting, we create new daily strips. Each day, Cybel encounters different situations that challenge their dual nature, from ethical dilemmas involving AI rights to personal struggles with their identity.

   The quick generation of dialogue and visuals through AI tools allows for rapid content creation, keeping the audience engaged with fresh, thought-provoking content.

Savings

Creating and publishing an original comic traditionally involves a detailed, time-consuming process spanning concept development, drawing, writing, layout, and distribution. This workflow demands a diverse skill set, including artistic talent, storytelling ability, and technical knowledge for digital publication.

Introducing AI into this process reshapes the landscape, offering significant savings in both time and money. Let's break down these savings through our example of creating a comic like "The Duality of Cybel."

Money

- **Creation:** Hiring professionals can be costly. Artists charge $100 to $300 per page, while writers charge $25 to $100 per page.
    - AI tools allow you to create directly, *saving thousands of dollars* and allowing anyone to explore their artistic side.

- **Software:** Professional software for drawing and layout (e.g., Adobe Creative Suite) requires a subscription, adding to the cost.
    - AI tools have free tiers or relatively low subscription fees, reducing the need for expensive professional software and the time to train on it.

- **Publication:** Self-publishing physical comics involves printing costs of hundreds to thousands of dollars. Digital publishing may have less upfront cost but still requires marketing expenses.
    - Digital platforms allow for free or low-cost publishing options, *eliminating the need for physical printing and distribution costs.*

Net Cost Savings: Simply put, AI makes creating a comic possible for most of us. While AI tools may cost us a few hundred dollars, we would have had to pay thousands to tens of thousands of dollars to commission someone to do this for us.

Time

- **Concept and Character Design:** This initial phase can take anywhere from a few days to weeks, depending on the complexity of the characters and the world-building required.
    - Creating a character concept can be *reduced to hours.*
- **Scriptwriting:** Crafting engaging dialogue and plotlines can take weeks, especially for longer comics or series.
    - Generate ideas and dialogue quickly, *turning weeks of work into hours.*
- **Artwork Creation:** Drawing, inking, and coloring a single page can take an artist eight to sixteen hours, depending on the detail. For a 22-page comic book, this translates to 176 to 352 hours of work.
    - Using AI image generation tools, *individual panels can be created in minutes,* dramatically reducing the time from concept to finished artwork.
- **Assembly and Lettering:** Layout and placing text in speech bubbles can take one to two hours per page, which amounts to twenty to forty hours for a standard-size comic.
    - Streamline the assembly process, *cutting the layout to a fraction of the time.*

- **Publishing:** Setting up for publication, whether online or in print, can add days to the timeline for formatting and submission.
    - Direct publishing on platforms like Webtoon simplifies the distribution process, *making it almost instantaneous* once the comic is ready.

Detailed Savings Example: "The Duality of Cybel"

Assuming a traditional approach might cost \$5,000 to \$10,000 and take four to six months for a 22-page comic book, AI tools can reduce the cost to under \$500 (accounting for subscriptions to AI services and any additional software needed for final touches). The time to create the same comic book can be reduced to a month or even a couple of weeks, depending on the complexity and the creator's familiarity with the AI tools.

Additionally, once you've invested in the tools, future comic books have minimal additional cost. You can make ten for the same price, whereas commissioning would require repeating all the original costs.

Tools and Tips

- **Character Creation:**

  Perchance or **GetMerlin** are wonderful for designing characters. Use **Artbreeder** for endless customization.[53]

[53] https://perchance.org; https://getmerlin.in; http://www.artbreeder.com

- **Story and Dialogue:**

  ShortlyAI generates story ideas and dialogue, and **Plot Generator** creates more comprehensive story elements.[54]

- **Artwork Generation:**

  DALL·E and **Craiyon** excel at creating detailed comic panels.[55]

- **Comic Assembly**:

  Use **Comic Draw** and **Canva** for layout and speech bubble placement.[56]

- **Publication**:

  Webtoon and **Tapas** are great platforms for publishing and community engagement.[57]

Freedom of Art

The advent of AI in comic creation marks a new era for storytellers and artists. It democratizes the comic-making process, allowing anyone with a story to tell to bring their visions to life, regardless of their artistic ability or financial resources. By harnessing these AI tools, creators can produce content faster, experiment with new ideas, and engage with audiences in ways previously unimaginable. AI isn't just changing how we create comics; it's expanding who can create them, making the world of comics more diverse and vibrant than ever.

---

54 https://shortlyai.com; http://www.plot-generator.org.uk

55 https://openai.com/dall-e; http://www.craiyon.com

56 https://plasq.com/apps/comicdraw; https://www.canva.com/

57 https://www.webtoons.com/en; https://tapas.io/

CHAPTER 20

AUTHOR A CHILDREN'S BOOK

Writing a fiction book is a journey of challenges, from developing a compelling plot to creating memorable characters. In recent years, Artificial Intelligence has become an invaluable ally in this creative endeavor, offering tools and solutions that streamline the writing process, enhance creativity, and even assist in designing captivating illustrations.

This chapter explores how AI can transform the ideation phase into a polished, ready-to-publish book, focusing on creating a children's book as a practical example.

Choose a Theme

Many of you likely have ideas for a book you'd want to write or some general themes that interest you. Of course, if you don't have any, you can begin at the idea generation stage, including as much or as little as you want in your prompt to a chatbot.

For this example, to demonstrate the power of personalization with AI and how it can profoundly impact your family's life, we will imagine a set of parents with a seven-year-old daughter, Mia. As is common, Mia faces fears that impede her daily life and worry her parents.

Addressing fears through creative storytelling can be therapeutic and empowering. Let's see how parents can use storytelling and illustration to help a child conquer fear, transforming a personal challenge into an engaging, instructive tale. The tools and process here could be used to write and publish any children's story.

Step-by-Step: The Process from Beginning to End

Here's an overview of the process, including how to collaborate with children and your AI tools to make it even more meaningful. AI's speed of work makes it a compelling and child-friendly integration partner.

1. Brainstorming Ideas

AI Tool: Chatbots such as ChatGPT or Claude.

Process: Use the chatbot to brainstorm ideas around the theme. You can ask for different plot lines, character descriptions, and potential story arcs.

- Example Prompt: *Suggest five different story ideas for a children's book about a child who is afraid of the dark.*
- Outcome: The AI will provide various plot ideas, such as ***a child overcoming their fear of the dark by***

befriending a Starfish who fears the darkness at the bottom of the ocean.

HINT: The mobile version of ChatGPT handles voice incredibly well. Talk to it like a friendly collaborator. Current versions allow you to speak to it naturally without waiting for it first to finish its speech. Let the child join in the process, expressing their opinion and making suggestions. In this instance, we will make Mia an active participant in the creation of the story.

2. Outlining the Story

AI Tool: Chatbots such as ChatGPT or Claude or a writing tool such as **Squibler**.

Process: Create a detailed outline of the story that was initially provided as one of the ideas.

- Example Prompt: *Help me outline a story where a child overcomes their fear of the dark by befriending a Starfish who fears the darkness at the bottom of the ocean. Name the child Mia, a human child who also shares a fear of the dark each night she tries to sleep.*
- Outcome: The AI will break down the story into chapters or sections, providing a structured approach to the narrative.

HINT: You can prompt the AI to make changes and alternative ideas for any section. You can also copy the outline into your preferred Word document program (Microsoft Word, Google Docs, etc) and make some edits before moving to the writing stage.

3. Writing the Story

AI Tool: Chatbots like ChatGPT or Claude or a writing tool like Sudowrite.

Process: Write the full text of the children's book.

- Example Prompt: *Based on this outline, write a 1000-word story about the Starfish named Bella and her human friend Mia, who helps them overcome their fears.*
- Paste the final edited outline into the text field.
- Outcome: The chatbot will generate a draft of the story.

HINT: You can request revisions or additions as needed to the chatbot or, like with the outline, edit it yourself and save a new document.

4. Editing the Story

AI Tool: Grammarly or Hemmingway App

Process: Edit the text for grammar, punctuation, and style.

- Paste the story into the appropriate fields of the AI tool to highlight errors and suggest improvements.
- Outcome: A polished and grammatically correct manuscript.

These tools may have a cost, so editing can be done the old-fashioned way by pasting it into a Word program and reading it through. Both Microsoft Word and

Google Docs have free proofreading tools to highlight areas of improvement.

5. Illustrating the Book

AI Tool: DALLE, Canva, or Artbreeder.

Process: Create illustrations for the book.

- Example Prompt: Generate illustrations for each page of the story *The Brave Little Starfish.*
    - Attach the most up-to-date version of the text with the prompt.
    - It is better to break the book into chapters and ask AI to illustrate it section by section.
- Outcome: AI will produce detailed images based on your descriptions. You can refine the prompts to match your vision for the book's artwork.

HINT: This is a great time to involve the child in the process if you haven't already. Mia could help choose colors for her character's clothing and suggest ideas for the scene settings or other details. Simply ask the AI to recreate the image with the new ideas. With Artbreeder, it's easy to provide material such as a photo of Mia and even the parents and "remix" them into the style of the story and characters, creating a true deep personalization of the experience.

6. **Combining Text and Illustrations**

   AI Tool: Canva

   Process: Use Canva to combine the text and illustrations into a cohesive book format.

   - Upload the text and images into Canva and use its design tools to create appealing page layouts.
   - Outcome: A professionally designed book layout ready for publishing.

7. **Self-Publishing the Book**

   AI Tool: Amazon Kindle Direct Publishing (KDP)

   Process: Publish the book using Amazon KDP.

   - Create an account on KDP, upload your book file, and follow the steps to publish.
   - Outcome: Your book will be available on Amazon in digital and print formats.

Personalized Narratives

Incorporating the child Mia into the narrative of *The Brave Little Starfish* not only customizes the experience but also provides her with a mirror to reflect her own journey of overcoming fear. Let's see how this works.

The Story Enhanced with Mia's Character

The starfish, Stella, ventures out each night to discover the ocean's wonders, but the darkness always scares her away.

One evening, Stella encounters a kind-hearted girl named Mia, a visitor from the land above who shares Stella's fear of the dark.

Stella and Mia immediately bond over their shared fear, and together, they embark on a nighttime adventure to conquer it. Mia tells Stella about the nightlights that keep her safe and warm back on land and wonders if the ocean might have its own nightlights.

As they explore, they encounter glowing jellyfish, bioluminescent algae, and other radiant sea creatures, all offering a soft glow to light their path. Each creature they meet shares wisdom about the beauty found in the dark, and slowly, Mia and Stella start to see the night in a new light.

Emotional Growth

Mia's character in the book serves as a guide and a companion to Stella, symbolizing the real Mia's own capacity for bravery and her quest to overcome her fears. This parallel story within a story empowers Mia, showing her that she can be both the hero in her own life and an inspiration to others.

Best of all, Mia has been included in the process, making the story truly hers and authentic. Sure, AI played a key role, but it supported every suggestion and change asked, re-writing and re-creating with infinite patience.

The process and book offer a fantastic development opportunity for children like Mia:

- Seeing herself in the story helps Mia to externalize her fear, making it something she can observe and address rather than an overwhelming internal experience.

- Creating the book becomes a therapeutic exercise, allowing Mia to control the narrative of her fear and ultimately rewrite her relationship with the dark.
- By the end of the book, Stella and Mia learn that while the dark can be scary, there's always light to be found, and they have the courage to seek it out.

Through *The Brave Little Starfish*, Mia's journey to bravery is immortalized in the pages of her personalized book, reminding her of her strength and the love of her family that supported her through her challenge.

AI's Vast Capabilities in Fiction Writing

An illustrated children's book only scratches the surface of AI's potential to be your writing collaborator. Whether fiction or nonfiction, graphic novel or epic trilogy, AI can be a powerful partner for your writing aspirations. These include:

1. Brainstorming and ideation
2. Genre-specific character and plot generation
3. Emotional tone analysis for a more profound narrative impact
4. Use of history, geography, science, and other subject matter incorporated into your story
5. Research and fact-checking
6. Illustrations and book design
7. Automated marketing copy for book promotion
8. Recommendations of reader preferences and trends

Savings

Money

- **Professional Writer:** Hiring one would cost thousands of dollars.
    - **Savings:** You save *thousands of dollars* even with monthly subscriptions of $10 to $20 on AI tools.

- **Illustrator:** Hiring an illustrator costs between $100 to $500 per illustration. For a 20-page book with one illustration per page, the price could be $2,000 to $10,000.
    - **Savings:** The subscription costs for AI tools are nominal and will save you *thousands of dollars.*

- **Designer:** Book layout design services might cost upwards of $500.
    - **Savings:** The free or low-cost AI alternatives *save you at least 80 percent.*

- **Publishing:** Self-publishing a book traditionally incurs costs for printing and distribution without a guarantee of sales.
    - AI-driven platforms offer print-on-demand services, reducing the upfront investment. This is variable depending on printing costs and how many you sell, but any level of print inventory puts thousands of dollars at risk.

Estimated Net Money Saved: *$2,500 or more,* not including potential savings from print-on-demand services and reduced need for inventory risk.

Time

- **Brainstorming and Ideation:** Without AI, brainstorming could take several sessions over days or weeks. AI can generate a multitude of ideas in minutes.
    - **Savings:** *About five hours.*
- **Writing:** A typical children's book could take a professional author over a hundred hours to write. With AI assistance, the writing time can be cut in half or more as AI can instantly provide suggestions and edits.
    - **Savings:** *fifty to seventy hours.*
- **Illustrations:** Hand-drawing illustrations or hiring an artist could take weeks. AI can generate initial illustrations in minutes, which can be refined as you want through "in-painting," a method of prompting the AI to change only portions of an image.
    - **Savings:** *About forty hours.*
- **Layout and Design:** Manually formatting a book could take a novice over twenty hours. Many AI tools offer templates and automated layout options that can be completed in under an hour.
    - **Savings:** *About nineteen hours.*

Net Time Saved: *114 to 134 hours.*

Opportunity

Like a comic, the truth is you'd be unlikely to write a book given the typical skill, costs, and time requirements. Children's books, with the additional graphic requirements, offer an even higher barrier for most people to try. Other types of books also have significant challenges of time, research, and skill, making them beyond the reach of most of us.

Further, in our example case, the minimal cost and immediacy of AI-generated content and the ability to rapidly iterate and refine the book collaboratively means that Mia and children like her can get their own personalized tool for overcoming fear in their hands much sooner and with much less financial investment than through traditional methods. It demonstrates that AI not only accelerates the creative process but also makes it far more accessible and possible.

Tools and Tips

- **Writing and Ideation**: **ChatGPT**, **Claude**, **Scrivener,** or **Sudowrite**[58]
- **Illustrations**: **DALL·E, Canva**
- **Editing: Grammarly, Hemmingway App**
- **Formatting and Printing: Vellum** can turn e-books into beautiful objects and enable you to print paperback versions if you desire the physical thing.[59]

A New Age of Creation

Incorporating AI into writing a fiction book, particularly a children's book, is not just about efficiency; it's about

[58] http://www.literatureandlatte.com/scrivener/overview; http://www.sudowrite.com

[59] https://vellum.pub

unlocking a new realm of creativity and possibility. From the spark of an idea to the final printed page, AI is a companion, guiding authors through each step, making the once-daunting task of book creation accessible and enjoyable.

The speed, adaptability, and personalization enabled by AI open new forms of creative collaboration, where writing becomes an invaluable tool for emotional development and exploration. As we look to the future, integrating AI into creative processes promises to bring forth a new era of storytelling where the only limit is our imagination.

SECTION VI

FUN AND FAMILY

CHAPTER 21

A PLAYLIST FOR ANYTHING

We've tackled home improvement, bettered our health and finances, made ourselves more productive, and cracked open our creative souls. In the coming chapters, we'll see how AI can also help us get more joy from living, beginning with something we universally love.

Music can evoke emotions, enhance experiences, and create lasting memories. It's an essential part of our lives, whether we are sharing time with friends, meditating on love, embarking on a road trip, diving into our favorite books, or reliving moments through pictures.

Personalized playlists amplify these experiences, making them even more unique and memorable. AI makes creating these tailored musical journeys accessible to everyone. It also allows us to explore the playlist in ways we may not have previously considered. Let's take the journey of constructing soundtracks for any experience conceivable.

Music and Imagination

Modern AI is built on information and data. When music began to get digitized over thirty years ago, along with the audio came lyrics, artist biographies, articles, interviews, and more. This enabled not only the music to be easily accessible but also the universe of emotion, experience, and storytelling that is evoked by each track. This has given LLMs vast training data to help them understand what each song means to people.

Let's explore how one app, Playlist AI, has turned this into a delightful, moving, and endlessly surprising addition to our musical experiences.

Step-by-Step: Call Out to the "Road" in Road Trip

You're planning a road trip, so you want road trip music. Yet what makes the trips exciting is that each stretch of road is different. The variety is as expansive as nature is on earth. So, how do you get just the right sound for the moment and place you are in? Here's how you can use PlaylistAI, an app that also has a ChatGPT extension, to create the perfect playlist.

AI Tool: Playlist AI

1. Identify the Experience

Prompt not simply for road trip music but for the specific time and place you are in. Example: *Driving along the Pacific Coast Highway south of Monterey. Breathtaking views of the ocean, cliffs, and redwoods. Curving roads, steep drops, beauty, and danger.*

2. Refine Your Playlist

Review the AI's track suggestions and (if you desire) modify them by specifying genres, artists, songs, or moods you want more or less of. AI might not know every suburban basement band that has ever played, but chances are it will know whoever or whatever you have ever listened to.

3. Finalize and Export

Once you are satisfied with the list, connect to your preferred music streaming platform, such as Spotify or Apple Music. The app will create the playlist and drop it into your library.

Let's explore additional examples of how AI can build playlists for scenarios other than the usual activities and moods.

Inspired by a Favorite Book

Scenario: You've just finished reading *The Night Circus* by Erin Morgenstern, and you're captivated by its magical atmosphere, vivid imagery, and the intricate relationship between Celia and Marco. You wish to extend the experience through music.

1. **Input Description**

   Describe the book's atmosphere, key themes, and emotions to PlaylistAI. Mention the blend of magic, romance, and the mysterious competition between the protagonists. You can also ask an AI chatbot to summarize the book and then use the summary to prompt for the playlist.

2. **Output**

   A list of songs that evoke a sense of wonder, magic, and the complex dynamics of love and rivalry, with instrumental tracks that convey the mystical and enchanting circus environment.

Based on an Artist's Work

Scenario: You're fascinated by Vincent Van Gogh's "Starry Night" and want a playlist that captures the emotional depth and beauty of the painting.

1. **Input Description**

   Describe the emotions and themes of "Starry Night"—the swirling skies, the vibrant colors, and the emotional turbulence Van Gogh was experiencing.

2. **Output**

   A selection of music that mirrors the painting's swirling patterns and emotional intensity, including tracks that reflect the tumultuous beauty of the night sky and Van Gogh's emotional state.

For a Personal Writing Piece

Scenario: You've written a piece of personal writing, such as a poem or short story, that's deeply meaningful to you, and you'd like a playlist that complements its themes and tone.

1. **Input Description**

   Share critical themes, the mood of the writing, and even specific, evocative lines.

2. **Output**

   It curates a selection of songs that resonate with the emotional landscape of your writing, enhancing the narrative and mood you've created with your words.

Additional Ways Playlist AI Can Be Used

1. **Thematic Playlists**
   - Create playlists based on specific themes or emotions portrayed in artwork, literature, or personal experiences.
2. **Event-Specific Playlists**
   - Generate music for events directly inspired by cultural or artistic milestones, combining historical context with contemporary tunes.
3. **Mood-Based Playlists**
   - Curate playlists that match the mood or atmosphere of a piece of art or literature, enhancing the overall experience of the work.
4. **Character or Story Playlists**
   - Develop playlists that reflect a character's journey in a book or film, using music to deepen the connection with the narrative.
5. **Fitness Playlists**
   - Tailor workout tunes to your exercise intensity.

6. **Study Playlists**
    - Find ambient music for studying or relaxation exactly how you want it.
7. **Photo Playlists**
    - Music based on a set of pictures can evoke the emotions captured in those moments.

These examples illustrate just a fraction of the potential applications for creating deeply personalized playlists. By leveraging AI's capacity to interpret and respond to a wide array of inputs, users can craft soundtracks uniquely tailored to their experiences, preferences, and the art that moves them.

Savings

Money

- **Maximized Subscription Value**: Most streaming services require a subscription. AI-enhanced playlist creation helps you discover more music that aligns with your tastes and moods, ensuring you get the most out of their subscription fees.
- **Reduced Subscription Need:** By efficiently using one platform to its fullest potential, you might feel less need to subscribe to multiple services to find the music you like, potentially saving on subscription costs.
- **Decrease of Unnecessary Purchases:** AI's ability to tailor music so closely to user preferences reduces the tendency to buy songs or albums that don't fully meet listening needs. AI's precision in curation can lead to

more satisfaction with the music discovered and less spending on music that doesn't fit.

Estimated Net Money Saved: Exact savings in dollars can vary widely based on individual spending habits and subscription models. Optimizing a single streaming service subscription (typically $10 to $20 per month) can negate the need for multiple subscriptions or music purchases, saving you *hundreds annually*.

Time

1. **Rapid Discovery and Compilation**: The traditional method of creating a personalized playlist involves manually searching for songs, listening to them to assess their fit, and compiling them. This process can easily take several hours, especially for those seeking to personalize their playlists for specific themes or experiences.
   - AI can *reduce this to minutes* by instantly analyzing vast music databases and user preferences to generate relevant playlists.
2. **Automated Curation**: AI can understand complex inputs, such as a book's mood, a painting's thematic elements, or a personal experience's vibe, and curate playlists accordingly.
   - This automation *saves hours* that would otherwise be spent in manual curation.
3. **Efficiency in Refinement**: Adjusting and refining playlists based on new discoveries or changing preferences is time-consuming. You might need to

manually redo much of the work to adjust a playlist's vibe or theme.

- AI allows for swift updates.

Net Time Saved: *three to five hours* for researching, sampling, and curating a playlist of fifty songs.

Tips and Tools

- **PlaylistAI GPT/App** creates playlists using *ChatGPT,* enabling playlists for, well, anything you can think of asking.[60]
- **Spotify's AI DJ** offers personalized music recommendations based on your listening history.[61]
- **Playlistable** allows you to select your mood, genre, artist, or song to create a Spotify playlist suited to your current feelings.[62]
- **Pandora** uses the Music Genome Project to curate playlists based on the musical traits of your favorite songs.[63]

Reimagining Our Music

AI-enhanced playlist creation is an example of how this new transformative technology opens up endless possibilities for customization. In this chapter, we discovered how

[60] https://playlistai.com

[61] https://www.spotify.com

[62] https://playlistable.io

[63] https://www.pandora.com

it can create the soundtrack to our lives so we're always in perfect harmony with our moments. In doing so, we find new artists, save time searching, and emotionally connect to elements of our lives and the world we value and treasure. With tools like PlaylistAI GPT and other innovative apps, everyone can effortlessly and beautifully become the curator of their musical universe.

CHAPTER 22

NO STRESS TRAVEL

We all look forward to vacations, but few of us like to plan them. We don't have the time, there are endless options and decisions to make, and managing costs is highly stressful. However, the advent of AI in travel planning has transformed this daunting task into an efficient, enjoyable process.

We'll delve into the world of AI-driven vacation planning, using a specific example of a family planning a trip to London and Paris to illustrate the step-by-step process and the significant benefits of time, money, and reduced anxiety. You're going on vacation; let's eliminate all the stress of planning it!

Being a Smart Traveler Starts with Planning

AI in vacation planning leverages machine learning, data analysis, and automation to offer personalized travel suggestions, streamline booking processes, and provide real-time travel assistance. This technology can handle everything from suggesting destinations to booking flights and hotels, planning itineraries, and even offering language support. Want to visit Europe? Let's go!

European Vacation

We'll call this the Harris Family Adventure. With two parents, two children, and the pets left at home, the Harrises decided to embark on a European journey, aiming to explore the historical marvels of London and the romantic streets of Paris. With a moderate budget and a desire to experience the best of both cities, they turn to AI for assistance.

Travel Planning Overview

A good sequence to follow would be:

1. Choose Destination
2. Itinerary Planning (with dates)
3. Flight Booking
4. Hotel Booking
5. Additional Bookings (Tours, Destinations, etc.)

Many AI-powered solutions are comprehensive and can go through all the steps, including direct links to complete reservations. We'll walk through each step to recommend specialty tools that might save you more or offer a more thorough category analysis.

Step-by-Step: Embrace Otherworldly Intelligence

1. Destination Insights

AI Tools: iPlan AI, ChatGPT, Curioso

Process: The Harris family knows where they want to go. If they didn't, then AI can help. **iPlan** is an excellent tool for this. Input your interests, time, budget, and suggestions are in your hands within seconds. Your preferred AI chatbot, such as ChatGPT, can also help guide you through the process. Begin the conversation as if talking to a friend about traveling somewhere, and within minutes, you'll have recommendations galore.

Curioso is a fun trip discovery app focused on road trips. Tell it where you want to begin and end, and it will offer a variety of road trips at different lengths of time and cost. If you have a few days and haven't given a break much thought, it's a great way to discover adventures within easy reach of home—no worry about flights and the costs and logistical issues that accompany them.

2. Itinerary Planning

AI Tools: AskLaya, Trip Planner AI, Wonderplan.AI, ChatGPT

Process: New AI travel planners merge conversational chat with travel planning, a significant leap forward for efficient and user-friendly itinerary building. Unlike some other specialized tools mentioned throughout the book, most AI travel tools do not require a subscription as affiliate fees support them through flight, hotel, or tour bookings (you are not required to use these sites for those bookings, and we'll recommend other specialized tools for these to help you get the best price).

Specific Example: The Harris Family told the **AskLayla** chat assistant they wanted a ten-day trip to London and Paris in April, and within seconds, they had a draft of a detailed daily plan.

- They refined the first draft to suit their interests, telling it to focus on more historical and outdoor activities like walking tours and fewer museums. Seconds later, another itinerary was delivered.
- Each day was divided into segments, the distance between sites was considered, and restaurant recommendations were made in the vicinity.
- As if conversing with a friendly travel agent who knew the location intimately, the Harrises refined their itinerary in less than thirty minutes.

3. Flights

AI Tools: Skyscanner, Hopper

Process: The Harrises turned to **Skyscanner** to find the best combination of low-cost flights and convenient travel times to match their itinerary dates. Skyscanner also has a fun feature if you're not sure where you want to go. Plug in your home airport and see where you can get on your budget for the dates you want to travel.

If you are flexible on dates or want to wait for the best fare, Hopper will track your dates and destinations and send alerts when prices change. It also uses AI to recommend the best time to book flights to get the best deals.

Flights can be quickly booked through Skyscanner, although once you find the flights of your choice, it is always good to check the airline website directly to ensure you are getting the best fare.

LOOK OUT! If you book through a travel site, check the rules around changes or cancellations. These can sometimes be more restrictive than when handling directly with an airline.

4. Accommodations

AI Tools: Trivago, Kayak

Process: Looking at lists of options can be overwhelming, but when it comes to hotels, a nice, clean list of different choices with cost, ratings, and location is extremely helpful. The Harrises turned to Trivago, which rapidly checks hundreds of booking sites to summon a well-organized list nearly instantly.

Specific Example: Armed with their itinerary, the Harrises had a good idea of where they wanted to stay. Trivago provided easily applied filters for each part of London and Paris, along with types of accommodations, ratings, and costs. They could even choose apartments or other alternative stays besides hotels. Accommodations were shown on a map of London alongside the list to make it even easier to focus on a set of hotels in a particular part of town.

TIP: These AI tools cover Booking.com and other popular internet sites, so you don't need to check all of them individually to take advantage of their best offers.

5. Tickets and Tours

AI Tools: AskLaya, Trip Planner AI, Wonderplan. AI—the same tools as for itineraries—will offer links to buy tickets or list for tours.

Process: Once the Harris's had their dates all booked, they could return to the itinerary and finalize specific tours or ticket times. Doing this before getting final flights and accommodations could have been a waste of time, especially if you have some flexibility in dates to get the best rates.

Specific Example: Before committing to tickets, the Harris family returned to AskLayla and drilled into the specifics of the itinerary. Using the AI chatbot with simple natural language, they asked about more rarely visited spots—hidden gems—for one afternoon and an outdoor excursion on another morning. Options for both came without hesitation.

Wanting to be sure they had a complete picture of options, they turned to Chathub and queried multiple AI chatbots with the prompt, "*What hidden gems in London would make our visit unique?*"

Each of the six AI assistants provided a list of ten or more special gardens, museums, and lesser-known parts of the city, each with a quick description. In a few minutes, the family picked out a couple more additions to their itinerary and returned to their AI planner, which slotted visits to the new destinations based on the best times given their location each day.

You could use the AI itinerary planning tools to complete all aspects, including flights and hotels, but better AI tools may be available. It's convenient, and pricing will be competitive, but for a costly, once-in-a-lifetime trip with the family, a little extra time spent is well worth it.

6. Local Transportation

AI Tools: Citymapper

Process: Let's jump ahead to the trip. There are some great AI tools to get the most out of your journey. The **Citymapper** app provides real-time updates on public transport, including buses, trains, the Tube in London, and the Metro lines in Paris. Citymapper's AI analyzes the quickest and most efficient routes, considering the location, desired destinations, and any current disruptions in the transportation network.

Traffic, breakdowns, or getting lost could create hours of delays, costing you valuable time and adding unnecessary stress to your vacation.

7. Language Assistance

AI Tools: Google Translate, ChatGPT

Process: The Harris family were not French speakers, so they turned to Google Translate for essential interactions. While the highly visited tourist sites have English signs and hosts, the Harrises liked to get off the beaten path. The app's AI-powered camera translation feature was handy, allowing them to instantly translate menus, signs, and instructions by simply pointing the camera at them.

Live conversation mode also helped them communicate fluently with locals, enriching their cultural experience and making their Paris visit more enjoyable. With a couple of taps, the app opens the microphone and translates what it hears in real-time.

Tip: While Google Translate has long been the go-to for language translation, ChatGPT 4o introduces a powerful real-time translation ability. With the mic open on the mobile version, it instantly recognizes and translates foreign language and can quickly translate your own speech into the language of your choice. Be sure to download this as part of your AI toolkit when on the road in any foreign country. Undoubtedly, Google Translate will continue to get significant upgrades to compete.

8. Cultural Enrichment

AI Tools: ChatGPT, Gemini

Process: Travel is often about learning, discovering, and enriching your mind and soul. AI chatbots now provide 24/7 access to information and insight about where you are and what you are experiencing at any time, any place in the world.

Specific Example: Whether a historical landmark like Kensington Palace or Notre Dame, a piece of art seen in the Louvre and an ancient relic in the British Museum, or a side street in either city with a mysterious monument or artwork, the Harris Family always had a tour guide in their pocket to query and provide answers and access more profound knowledge. When it comes to historical, geographical, and cultural questions, the intelligence of all the world's experts is readily at hand.

The Outcome

The Harris family maximized their vacation enjoyment by leveraging these AI tools while minimizing planning stress. They explored iconic sights, enjoyed local cuisine without the fear of language barriers, and navigated each city with ease. Their trip was not only about visiting two of the world's most beautiful cities but also about experiencing the blend of history, culture, and modern convenience through the lens of AI-enhanced travel.

AI Assistance in Travel Planning

Beyond the example provided, AI can assist in nearly every aspect of travel planning.

1. **Flight Booking**
   - Predicts flight price changes and suggests the best times to buy tickets.
2. **Accommodation Selection**
   - Matches you with accommodations based on preferences and budget, including hotels, hostels, and vacation rentals.
3. **Itinerary Customization**
   - Tailors daily itineraries based on interests, travel pace, and local events, ensuring a personalized experience.

4. **Packing Assistance**
    - Suggests packing lists tailored to the destination's weather forecast and planned activities.
5. **Dining Recommendations**
    - Analyzes reviews and preferences to advise on restaurants and cafes.
6. **Activity and Event Booking**
    - Recommends books, tours, museum visits, and events, optimizing for preferences and schedule.
7. **Safety Advice**
    - Monitors travel advisories, health alerts, and local emergency services information, offering real-time advice.
8. **Transportation Options**
    - Suggests the best local transportation options, including public transit, ridesharing, and bike rentals.
9. **Travel Insurance**
    - Recommends insurance options based on the trip's details and traveler's needs.
10. **Cultural Etiquette**
    - Provides insights into local customs, etiquette, and helpful phrases to ensure respectful interaction.
11. **Sustainability**
    - Offers advice on making travel more eco-friendly, including sustainable accommodation and activities.

12. **Virtual Reality Previews**
    - AI-driven VR experiences allow travelers to explore destinations and accommodations before booking.
13. **Expense Tracking**
    - Helps manage travel budgets and track expenses in real-time.

Savings

Time

- **Research and Planning**: Manual research and planning for a vacation to a new and unknown location takes extensive time.
    - **Savings:** *twenty to forty hours*
- **Smart Itineraries:** AI-powered itineraries organize destinations based on their vicinity to other parts of the itinerary.
    - **Savings:** *Hours* of travel time while maximizing the amount seen.
- **Travel Time**: Optimized based on AI analysis of the real-time options and conditions.
    - **Savings:** *Hours* in potential traffic jams or travel hold-ups.

Net Time Saved: *twenty-five hours or more* before and during the trip.

Money

- **Flights and Accommodations**: Predictive pricing and AI-powered comparative searches save money on these expenses, allowing the budget to be used elsewhere.
    - **Savings:** *twenty percent less for these essential costs*
- **Detailed Pre-Planning**: Spontaneity can be fun, but it can also lead to costs slipping out of control. Travel costs become inflated when you are forced to negotiate prices on the spot or find restaurants and accommodations when under pressure and exhausted.
    - **Savings:** Optimizes experience to stay within budget, avoiding greater than twenty percent cost escalation.

Estimated Net Saved Money: *$1000 to $2000* on a family trip of $10,000, and avoidance of going over budget.

Stress

- **Planning Process**: Family travel is exciting but can also create a lot of anxiety and tension. The streamlined planning process and real-time AI assistance during the trip significantly reduced stress, making the journey more enjoyable and a true *vacation*.

Tools and Tips

Largely free and plentiful, AI is a powerful ally but can also overwhelm in this category. The tools mentioned above and here are only a sliver of what is available. These are a good start, but new ones may emerge, and your past favorite web travel tools will all likely be getting an AI boost soon if they haven't already.

- **Choosing Destinations**:
  a. **iPlan AI** and **ChatGPT** are great for getting ideas.[64]
  b. Use **Skyscanner** (Explore Mode) to pick a flight destination based on your home city and budget.[65]
  c. **Curioso** can help select a spontaneous road trip from your home city.[66]
- **AI Travel Planners**
  d. **AskLaya** and **ChatGPT** can be conversation planning assistants.[67]
  e. **Trip Planner AI** and **Wonderplan AI** have simple choose-and-find interfaces.[68]
- **Flights**
  f. **Skyscanner** searches the nooks and crannies of the internet for the best price, and **Hopper** has excellent price prediction and alerts for your dates and destination.[69]

---

[64] https://iplan.ai;

[65] https://www.skyscanner.net

[66] https://curiosio.com/

[67] https://asklaya.com

[68] https://tripplannerai.com; https://wonderplan.ai

[69] https://www.hopper.com

- **Accommodations**
  g. **Trivago** and **Kayak** are potent tools with friendly interfaces for finding the best deals. Explicit ratings and benefits and integrated maps accompany them.[70]
  h. **Staypia** and **Eddy Travel** both focus on budget stays.[71]
  i. **HotelTonight** and **LastMinuteTravel** are good tools when you need a place in a pinch.[72]
- **Reviews and Recommendations**
  j. If you want to take the extra step in researching your accommodations for Things to Do, **TripAdvisor** and **Yelp** are still the go-to for activities and dining options.[73]
- **Language**
  k. **Google Translate** remains an app for traveling to countries where you don't speak the language, but **ChatGPT** mobile offers remarkable real-time translation.
  l. Want to learn the language? **Duolingo** and **Rosetta Stone** are solid choices.[74]

---

[70] https://www.trivago.com; https://www.kayak.com

[71] https://www.staypia.com; https://eddytravel.com

[72] https://www.hoteltonight.com; https://www.lastminutetravel.com

[73] https://www.tripadvisor.com; https://www.yelp.com

[74] https://www.duolingo.com; https://www.rosettastone.com

- **Budgeting**
  m. **TravelSpend** and **Tripcoin** are excellent for tracking travel-specific expenses and budget management.[75]
- **Cultural and Historical Enrichment**
  n. Use **ChatGPT**, **Gemini**, or your favorite go-to AI chatbot.

Increase the Possibilities, Decrease the Worry

AI transforms vacation planning from a stressful task into a streamlined, enjoyable process. By leveraging AI tools, the Harris family maximized their London and Paris adventures and saved time, money, and stress. AI further enhanced their travel by bridging language barriers and providing insights into all the incredible places they saw and people they met.

Travel is an irreplaceable human experience from which everyone can gain immense joy and enrichment. The possibilities are endless for how AI can supplement your journey with access to better planning and budgeting tools before you leave, as well as free, powerful companion applications that will make your trips truly unforgettable.

[75] https://www.travelspend.com; https://tripcoinapp.com

CHAPTER 23

A PHOTO AND VIDEO EVOLUTION

In the digital age, where photos and videos are continuously captured, the challenge often lies not in recording memories but in organizing and presenting them in meaningful ways. AI steps into this realm with a solution that transcends the exhausting effort of manual sorting and editing. By harnessing the power of AI, you can effortlessly sift through vast collections of digital media to create themed compilations that tell a story, evoke emotions, and celebrate the moments you don't want to lose.

AI and the Great Organization

My friend Oliver wanted to create a special video for his daughter Madison, who was celebrating her tenth birthday. He aimed to compile a decade's memories into a heartfelt video. Madison's whole life had the benefit of occurring after the introduction of the smartphone, with much of her life's special moments captured so they could be easily revisited and not forgotten.

However, the ease of recording Madison's childhood presented an inevitable challenge for Oliver: tens of thousands of photos and videos scattered across various devices with little more than a timestamp. Oliver was disheartened.

Step-by-Step: Celebrating Life's Moments

Let's dive into a step-by-step guide that uses accessible and user-friendly tools to craft a memorable birthday video compilation. For this effort, **Google Photos** and **Clipchamp** provide a free tool powered by AI to deliver incredible content with speed and personalization.

1. Gather and Organize Your Media

Tool: Google Photos

Process: Oliver gathered all his photos into a folder on his computer. One look at the multitude of pictures with no clear labels or organization made him shudder. How would he ever get a video together on time—his daughter's birthday was in two days? He uploaded them all to Google Photos with one swipe of his cursor.

A. If you already have a Google Account like most of us, Photos will be available through the Google Apps menu.

2. Create an Album

Tool: Google Photos

Process: Oliver clicked Albums and New Album and named it "Madison's 10th Birthday Celebration."

3. Select the Media

Tool: Google Photos

Action: This is where the magic of AI will astonish you. Once the photos were uploaded and without further organization effort, Oliver could organize the images using simple keyword searches.

A. He plugged in "birthday" to gather images from Madison's past birthdays. Instantly, birthday images appeared, and he clicked and added the ones he desired to the new album.

B. He plugged in "cat" to get images of Madison's dear pet.

C. He plugged in "ballet," "theater," "skating," and "beach," and each time images from Madison's life appeared. Oliver could select whichever ones he wished for the album.

D. Oliver did all this without having tagged any of the images. The only thing the AI couldn't know was the names of the people. However, Oliver could choose a person and label them, after which the AI recognized all the images with that face.

E. Because Madison changed from a baby to a ten-year-old just like any child, AI needed a little help with a few "Madisons"—Baby Madison, Preschool Madison, and Grade School Madison covered it!

F. In a few minutes, Oliver added more labels for other family members, and then he could search for a person in any place with any object. After tagging all the photos he wanted for the video, he downloaded the album.

Hint: Once you have an album of images and videos to include, I recommend ordering how you think you'd want them to appear in a video. You can constantly re-arrange any material at the editing stage, but getting the first draft closer to the final order is helpful. To do this, add numbers, starting with "1", at the beginning of the filenames. If you label 100 pieces of media in order from 1 to 100, this is the order in which they will first appear in your video.

4. Create the Video Compilation

Tool: Clipchamp

Process: AI can sometimes feel like a bit of a cheat. Should things be so easy and so quick? When Oliver uploaded his photos to Clipchamp, which offers a robust free tier (and premium tiers if you do this often), he could complete a video with music in just a few minutes.

Of course, there are numerous editing points in which you can dive in deeper.

5. Review, Edit, Export

Tool: Clipchamp

Action: Oliver initially let the Clipchamp AI construct a video. He then used the "Edit" function to add his soundtrack, rearrange some of the images and add additional effects.

- Enough editing tools are in the free tier to complete a well-tuned video.
- Once done, you can export a 1080HD mp4.

This is AI's compelling opportunity. It can help you complete something quickly and support your interest in collaborating more fully, where some extra effort pays off with accelerated results.

6. Share Your Masterpiece

Tool: Google Photos

Action: You can share the MP4 directly from your social media, but videos can be large. Reloading it into Google Photos provides an easy way to share it with a link for anyone to enjoy. It also maintains itself in a cloud storage space that is accessible for presentation from anywhere.

The Insanity Metric

The whole process, from gathering scattered digital piles across devices to completing an exceptional, highly personalized ten-year birthday video spanning his child's life, took Oliver less than an hour! Such an effort would have usually been a twenty-five to thirty-hour slog with additional expertise to manage editing tools.

This is the exponential efficiency of AI at its peak, unlocking a realm of creative opportunity for anybody who takes pictures. In the age of the smartphone, I'd say that's pretty much every one of us. As is often the case with AI, accessibility is also extremely high with simple interfaces, suitable for anyone aged eight to eighty-eight who wants to construct their own videos.

AI in Personal Media

Just as picture moments are endless, so are the occasions to pull together and share them. Here are a few categories to consider, all quickly possible with the help of AI:

1. Create holiday and vacation highlight reels.
2. Compile "year in review" videos showcasing key moments.
3. Generate themed compilations for anniversaries, graduations, and other milestones.
4. Offer personalized movie creations from everyday moments.
5. Enable easy sharing and storing of compilations in the cloud.

Savings

Time

This is the monumental savings category.

- **Sorting Photos:** Manually going through thousands of photos could take well over ten hours.
    - **Savings:** AI reduces it to *minutes.*
- **Editing:** Making a ten-minute video can take twenty hours, and that's if you know how to use the editing tools. Learning them will take quite a few hours more.
    - **Savings:** *Many hours* are reduced to as little as a few clicks. With additional editing, the effort may be up to an hour, *reduced from over twenty hours.*

Net Time Saved: Thirty-plus *hours*, and significantly, the skills requirement has been eliminated, so it's easy for anyone to do.

Money

AI tools in media creation can reduce the cost for those who would instead turn to professionals to create a video for a special occasion rather than do it themselves.

- Professionals would cost a *minimum of hundreds* and, more likely, *thousands of dollars.*
- Working with a professional still requires a considerable amount of effort since you'll need to guide the editor, who will not know the context of the images or the family members. It's difficult for a stranger to compile something personal.

- Even if you use the tools suggested here for free, there will likely be some minor costs for cloud storage if you aren't already on a plan. This can range from $2 to $10 per month, depending on how much media you have.

Tools and Tips

- **Google Photos** offers facial recognition and can automatically create themed compilations.[76]
- **Clipchamp** utilizes AI to quickly craft videos from selected photos and clips based on the desired mood or theme and offers easy-to-use editing tools to modify.[77]
- **Adobe Express** is an alternative to Clipchamp. It's a little less intuitive and requires a bit more editor savvy, but it provides features such as an AI-powered option to select the best moments from video clips. It's also a good place to start if you want to get more serious about editing videos and prepare for a step up into the premium Adobe Premiere editing tool.[78]
- **Google Highlights** can help if you are extremely pressed for time. Google Photos allows you to make a "highlight reel" of photos and videos with a simple search and add function. There aren't many editing capabilities, but like the "memory clips" that often appear in our social media feeds, a

[76] https://photos.google.com

[77] https://www.clipchamp.com

[78] https://www.adobe.com

simple presentation provides a powerful emotional experience.

Making Memories Forever

The advent of AI in personal media creation opens a world of possibilities for celebrating life's special moments. Through intuitive apps and platforms, memories that might remain hidden in digital storage can now be brought to life, telling stories that resonate with emotion and nostalgia.

Oliver's experience creating a video for Madison's birthday exemplifies the power of AI to simplify the video compilation process and enrich our lives by preserving our memories in dynamic, compelling ways. Our past can now be regularly present, maintaining deep connections to the experiences and the people we love.

CHAPTER 24

FAMILY GAMETIME

Games have become highly mobile, and we all have at least a few on our phones. What has been gained in individual fun has been a loss for more socially shared game time. This can also exclude younger members of the family who don't have a phone yet.

Traditional family game nights must be planned and arranged, which makes them infrequent. You might want to enjoy game time together spontaneously, but you don't have the game or the correct number of people ready. Further, games can be played in various environments as a tool of education, mental challenge, and creativity.

AI enables all these possibilities, making playing games possible anytime, anywhere and turning it into a deeper and more educational experience. AI tools can help you design custom games, set rules, and involve everyone in creative problem-solving activities. This fosters positive engagement and strengthens family bonds.

Let's consider a couple of scenarios in which you might be together as a family and have an opportunity to turn dead time into fun, quality playtime.

Step-by-Step: Games for a Stormy Day

AI Tool: ChatGPT

Process: Let's start with a stormy day. As gusting winds make the pouring rain or falling snow extra treacherous and everyone's plans are canceled, people often retreat to their own spaces or get restless and grumpy. Here's how to use AI to play a social game on the spot and make it more custom to your ideas and creativity. Pop open ChatGPT on your phone or computer and dive in:

Prompt: *What games can we play right now as a family with your help?*

("I" is the AI chatbot)

1. 20 Questions

- I think of an object, animal, or person.
- Others take turns asking yes/no questions to guess what it is.
- After 20 questions, if no one guesses, I reveal the answer.

2. Story Building

- Start a story with one sentence.
- Each family member adds a sentence to continue the story.

- We can take turns, and I can provide sentences to keep the story going.

3. Trivia Quiz

- I can ask trivia questions on various topics.
- Family members take turns answering.
- Keep score to see who gets the most correct answers.

4. Charades (Text-based)

- One person thinks of a word or phrase.
- I can provide clues without directly saying the word.
- Others guess based on the clues.

5. Riddles

- I can provide riddles, and everyone can try to solve them.
- You can take turns guessing or work together.

6. Word Association

- One person says a word.
- The next person says a word that is related to the previous word.
- Keep going in a circle. I can provide words to keep the game going.

7. Would You Rather?

- I can ask "Would you rather" questions.
- Each person takes turns answering and explaining their choice.

8. Alphabet Game

- Choose a category (e.g., animals, foods).
- Take turns naming something in that category, starting with each letter of the alphabet.
- I can help by providing words if someone gets stuck.

9. Two Truths and a Lie

- Each person says three statements about themselves: two true, one false.
- Others guess which one is the lie.
- I can participate by sharing interesting facts and lies.

10. Memory Game

- I list a series of items or words.
- Players take turns repeating the list and adding one new item.
- The game continues until someone forgets an item. I can keep track of the list for you.

Depending on your children's age, any of these could be fun. They could also be customized for specific themes: trivia about cats, riddles about food, or story-building about cities in the clouds. You can request anything from your AI, as can your kids, which adds a lot of fun.

Let's say you want something more in the mood for this stormy day.

Unlocking Mystery and Adventure

Family: *We'd like to play an interactive mystery where we must find the culprit of a crime.*

AI: That sounds like a lot of fun; let's start by choosing a setting. Do you want it to be:

> ***A haunted mansion***
> ***A small village***
> ***A cruise ship***
> ***A school***
> ***A museum***

Family: *A haunted mansion.*

AI: **Perfect. Now, let's each choose a character. Here are the four roles:**

> ***Detective Johnson is a seasoned investigator with a sharp mind and a keen eye for detail.***
>
> ***Historian Laura has spent years studying the mansion's history and knows all its secrets.***
>
> ***Maurice, the Paranormal Investigator, always carries a ghost detector and has a knack for sensing spirits.***
>
> ***Jamie, the Tech Expert, can hack into security systems and find hidden files.***
>
> ***When you have each chosen a role, ask me to start the story. You can also make up your role.***

Family (after choosing roles): *We're ready to start the mystery.*

AI: Wonderful. A priceless artifact was stolen from the mansion last night. Only a few people were inside: the caretaker, a visiting professor, the gardener, and a ghost tour guide. The mansion is locked from the inside, so the thief is still here. What would you like to do first? Declare your character name when you ask a question.

So, each family member interrogates the suspects, investigates the house and grounds, and discusses the evidence. The AI customizes its response depending on the role and helps keep track of the information.

Like a good game host, the AI can recount anything previously said or discovered, provide hints, and expand on details. In many ways, the participants create the story through the questions they ask and the actions they take. They can even add traits or physical details to their character by declaring them to the AI, which doesn't miss a beat.

Like any good mystery, it ends with the key beat of an accusation against a suspect. Your family can play this competitively or collaboratively, whichever way they like.

Alternative Ways to Play

There are many approaches to playing mystery games with your AI chatbot.

- Everyone but one person leaves the room. That person creates the mystery with help from the chatbot, and the others then play it.
- The family helps build the mystery, defining their own characters, setting, and suspects. Once they've done this, the chatbot puts it together as a story filled with dialog and drama.

- The family roleplays all the suspects, not even knowing if they are the thief!
- Use DALL-E with Chat GPT or Gemini's picture generator to render images of the scene and even the evidence. Customize your characters with physical characteristics, clothing, and accessories.

The AI chatbot will adapt to whatever rules, story changes, or twists you want to throw at it, making the experience highly interactive, surprising, and personal.

Games for the Family Road Trip

Road trips can be a fantastic way for families to bond and explore new places together. However, long hours in the car can sometimes lead to restlessness and boredom, especially for kids.

AI can transform this travel time into a fun and educational adventure by helping you create and play a series of games tailored to your journey. These can keep everyone entertained and engaged with the places you are traveling to and through, making your road trip an unforgettable experience.

Setting the Scene

Your family is embarking on a road trip from San Francisco to Los Angeles. You want to make the journey enjoyable and interactive for your two kids, we'll call them Emma (10) and Liam (8). Here's how to use AI to create games that entertain everyone during the drive.

Game 1: Trivia

AI Tool: ChatGPT

1. **Preparation**

   Before the trip, the family generates trivia questions about the cities and landmarks they will visit. This helps everyone learn exciting facts about their route.

   - Prompt: *Can you generate trivia questions about San Francisco, Los Angeles, and famous landmarks along the way, such as the Golden Gate Bridge, Monterey, and the Hearst Castle?*

2. **Playing the Game**

   As they drive, the parents ask the trivia questions. Each correct answer earns a point, and the person with the most points wins a small prize by the time they reach their destination.

   - Example Question: "What year was the Golden Gate Bridge completed?" (Answer: 1937)

Game 2: Storytelling Adventure

AI Tool: AI Dungeon

1. **Story Setup**

   Using **AI Dungeon**, the family creates a collaborative story. The AI helps guide the narrative based on each family member's input.

   - Starting Prompt: *We are on a road trip and discover a hidden treasure map at our first stop. What happens next?*

2. **Playing the Game**

   Each family member takes turns adding to the story, with the AI providing suggestions and plot twists. This keeps everyone engaged and sparks creativity.

   - Example Plot Twist: ***As you follow the map, you encounter a mysterious stranger who offers to help. What do you do?***

Game 3: Scavenger Hunt

AI Tool: DALL-E or Gemini

1. **Preparation**

   The parents list items and landmarks to find along the way. The AI generates images and descriptions for each item.

   - Prompt: *Generate a list of fun items and landmarks to spot during our road trip, including images and descriptions.*

2. **Playing the Game**

   The kids receive a digital scavenger hunt list. They earn points by spotting and photographing items and landmarks.

   - Example Item: *A red barn along Highway 1.*

Game 4: Road Trip Bingo

AI Tool: Canva

1. **Creating Bingo Cards**

   The family creates custom bingo cards featuring things they might see on their journey (e.g., a police car, a beach, a cow).

   - Prompt: *Create a bingo card with items we might see on a road trip from San Francisco to Los Angeles.*

2. **Playing the Game**

   Each family member has a bingo card, and they mark off items as they see them. The first person to get five in a row wins.

   - Example Item: *An ice cream truck.*

Game 5: Language Learning

AI Tool: Duolingo

1. **Language Practice**

   Using AI language learning apps like **Duolingo**, the family practices Spanish together, learning words related to their trip.

   - Prompt: *Practice Spanish vocabulary related to travel and landmarks.*

2. **Playing the Game**

   The family challenges each other to use new Spanish words they learn during the trip.

   - Example Challenge: "Can you order food at our next stop in Spanish?"

Even More Games

In addition to the above, the complete list of ten games that ChatGPT gave us for the stormy day could also be great road trip games customized for the places you are in. There are countless fun facts and insights to be discovered in every spot you visit.

All the Ways to Enhance Family Game Time

AI can deepen and expand how we play together. Its features include:

1. **Game Design**
   - AI creates and customizes board games, card games, or digital games.
2. **Rule Generation**
   - It can generate and modify rules to suit different preferences and ages.
3. **Storytelling**
   - Develop interactive storytelling games where AI generates the narrative based on family members' input.
4. **Puzzle Creation**
   - Design custom puzzles, including crosswords, Sudoku, or logic puzzles.
5. **Language Learning**
   - Create language-based games to help kids learn new languages through interactive play.

6. **Art and Music**
    - Use AI to generate artwork and music for games, adding a creative dimension to the experience.
7. **Virtual Assistants**
    - Integrate virtual assistants to guide gameplay, provide hints, and track scores.

Savings

Money

- **Cost-Effective**: Many AI tools offer free or low-cost solutions, eliminating the need for expensive game purchases.
- **Reuse and Modify**: Custom games created with AI can be reused and modified, providing long-term value.

Time

- **Design Efficiency**: Creating games with AI takes minutes compared to hours of manual effort.
- **Instant Content**: AI can generate game content instantly, reducing preparation time.

Additional Benefits

- **Educational Value**: Engages children in learning activities through play.
- **Flexibility**: Allows for game time anywhere, fitting into busy schedules.

Tools and Tips

- **ChatGPT** or **Gemini** for brainstorming ideas and generating game content.
- **DALL-E** (in ChatGPT) for creating custom illustrations and game art.
- **AI Dungeon** for creating interactive storytelling games.[79]
- **Quizlet** for generating flashcards and trivia questions.[80]
- **Canva** for easy creation of bingo cards or other template game types.

Playing Games with a Personal Twist

AI is a powerful tool that can transform family game time, making it more interactive, creative, and accessible. Whether at home or on the go, AI allows you to design and customize games that cater to your family's interests and needs. You can foster creativity, problem-solving, and positive engagement, turning any moment into a memorable family experience. Embrace the potential of AI to bring your family closer and make game time an exciting part of your everyday life.

---

[79] https://play.aidungeon.io

[80] https://quizlet.com

SECTION VII

ACCESSIBILITY

CHAPTER 25

VOICE POWER ANYTHING

Accessibility in digital content is crucial for ensuring that everyone, regardless of physical abilities, can access, understand, and interact with technology. AI has become a powerful tool in breaking down barriers, particularly for those with vision or reading challenges. By integrating AI-driven voice-over capabilities and improving its ability to understand speech, we can transform digital content accessibility, making digital spaces more inclusive and user-friendly.

AI Voice is For Real, Finally

Text-to-speech (TTS) functionality has existed for decades. However, the robotic monotone that came to define it made the experience essentially unbearable. It was hardly a barrier-breaking experience that could allow those with reading difficulties, older adults, and others to access the world of digital publishing. Human attention and comprehension require more.

For what seemed like forever, training voice models was slow and expensive, and even then, it didn't produce a quality text-to-speech experience. It may have been okay for customer service or PSA announcements, but not for interactions that required a natural human touch. As we saw in the chapter about voice assistants and speech recognition, AI has revolutionized this space, making voice models rapidly and cheaply from small samples.

The results are astonishing and carry over into all forms of TTS, enabling the vast world of digital content to be genuinely available and digestible by anyone, regardless of their vision or reading capabilities. This is not only for text but also for video, where translation combined with voice models allows dialog in any language to be available in the same tone and voice as the original speaker.

Step-by-Step: Re-Discovering a Lost World

Meet Leonard, who, in his late seventies, found reading difficult. A voracious content consumer, Leonard has turned to audiobooks and podcasts but remains cut off from newspapers, magazines, and much of the World Wide Web. Retired and generally very healthy, Leonard had hoped he would enjoy the offerings of digitally published material for many more years. His condition has him feeling depressed and isolated.

1. Identifying the Need

Leonard's main challenge is accessing text-based content like email, newspapers, articles, and websites. He needs a tool to read text aloud to consume information audibly. Traditional text-to-speech was hard to focus on for very long.

2. **Choosing the Right AI Tool**

   With help from his family, Leonard opts for the AI-powered text-to-speech app **Speechify**. This app uses advanced AI algorithms to convert written text into natural-sounding speech, and it supports various formats, including PDFs, Word documents, and web pages.

3. **Setting Up the Tool**

   Speechify is installed on his smartphone and tablet. The settings are customized to suit his preferences, including selecting a preferred voice and reading speed and enabling Spanish, German, and English. Leonard is multilingual and enjoys Spanish and German material in his native tongue.

4. **Using the App for Personal Accounts and Web Browsing**

   Leonard can connect his email and cloud storage accounts (such as Google Drive) to access any of them in the app to listen to. He can copy the URL for any website and browse the curated material Speechify provides. He can bookmark his favorite sites to return to each day.

5. **Using the App for Reading**

   Leonard can open and listen to eBooks such as Kindle in the app. The AI-powered voice reads the text aloud, providing clear and accurate pronunciation. He can pause, rewind, or fast-forward the reading as needed.

6. Integrating AI Voice Assistants

Leonard's family also sets him up with an AI voice assistant like Google Assistant or Amazon Alexa. These assistants help him with scheduling and setting reminders.

7. Leveraging Optical Character Recognition (OCR)

Leonard can use his Speechify app on print materials. All he needs to do is take a picture of the page he wants to read and import it into Speechify. The app converts the image into digital text using AI-based OCR, which is then read aloud.

Comprehensive List of AI-Enhanced Accessibility

AI assists in the following ways:

1. **TTS Technology**
   - Converts written text into spoken words, allowing visually impaired users to consume digital content.
2. **OCR**
   - Transforms printed text into digital format, which can be read aloud or converted into Braille.
3. **Speech Recognition**
   - Enables voice commands for controlling devices, dictating text, and navigating applications.

4. **AI-Powered Screen Readers**
    - Provides audio descriptions of on-screen content, including text, images, and buttons.
5. **Language Translation**
    - Converts content into different languages, helping non-native speakers and those with reading difficulties.
6. **Personalized Learning Assistants**
    - Offers tailored educational content and assistance for individuals with learning disabilities.
7. **Voice-Controlled Smart Home Devices**
    - Enhances independence by allowing users to control home appliances through voice commands.
8. **Augmented Reality (AR)**
    - Assists with spatial awareness and navigation for visually impaired individuals.
9. **Real-Time Subtitles**
    - Provides instant captions for live audio or video, aiding those with hearing impairments.
10. **Emotion Detection**
    - Identifies and responds to user emotions, offering appropriate support or adjustments in communication.

Savings

Money

- **Affordable Solutions:** Many AI apps are free or low-cost, making them accessible to a broad audience. Subscriptions run $10 to $15 per month.
- **Reduced Need for Specialized Equipment:** AI-driven apps eliminate the need for dedicated screen readers or traditional software.
- **Lower Healthcare Costs:** Enhanced independence and accessibility reduce the need for frequent assistance, potentially lowering healthcare and support costs.

Estimated Net Money Saved: Apps and voice assistants will be an incremental cost, but they're highly affordable and allow access to so much content that newspaper and magazine subscriptions, etc., may become redundant or unnecessary. The price relative to time spent and the corresponding value is very high.

Time

- **Instant Access to Information:** AI tools like TTS provide immediate access to written content, saving hours spent on manual reading or waiting for human assistance.
- **Efficient Navigation:** Voice assistants streamline daily tasks, reducing the time required for manual input, information recovery, and physical movement to collect it.

Net Time Saved: *One to two hours saved daily* for waiting for assistance or slow access to information.

Tools and Tips

- **Speechify** subscription provides numerous voice models that enable natural-sounding speech. You can even train your own voice model.[81]
- **Voice Dream Reader** is an alternative to Speechify, but only for iOS.[82]
- **Seeing AI** is a free app from Microsoft that provides TTS and descriptions of the environment, products, and other aspects of regular sight. It assists those with severe vision limitations and is a powerful display of AI that detects and deciphers objects as well as text.[83]
- **Google Assistant** and **Amazon Alexa** are used in software and devices to provide robust voice recognition and conversation to access information, control devices, and support communications and scheduling.[84]

Accessibility Has Truly Arrived

AI is revolutionizing accessibility, making digital content more inclusive and user-friendly. These tools offer practical solutions for individuals with vision or reading issues and

81 https://speechify.com

82 https://www.voicedream.com/

83 https://www.microsoft.com/en-us/seeing-ai

84 https://assistant.google.com; http://www.amazon.com/alexa/

empower users to manage devices and software with only their voice.

By leveraging AI, we can significantly enhance independence, reduce costs, and improve the overall quality of life for those with accessibility needs. Embracing these technologies provides invaluable benefits to individuals and promotes a more inclusive society.

THE TRANSFORMATIVE POWER OF AI IN EVERYDAY LIFE

As we conclude this journey into artificial intelligence, it's inspiring to see how AI extends far beyond science fiction and grand technological feats. From curing diseases and combating climate change to improving our daily routines, AI offers a wealth of benefits that save time and money and enhance our personal growth. This book's insights illustrate how profoundly AI can transform our lives, paving the way for a brighter, more efficient future.

Global Impact of AI

We are in a world where doctors can diagnose diseases early and personalize treatment plans for each patient. This isn't a distant dream but a reality that AI is making possible. AI-powered imaging tools, which use advanced algorithms to analyze medical images, can detect cancerous cells in their infancy, giving patients a better chance at recovery. The impact on global healthcare is nothing short of revolutionary.

Environmental conservation also sees AI playing a pivotal role. Predictive analytics, which use historical data to anticipate future events, help forecast extreme weather events, allowing for better preparedness and response. Smart grids balance energy loads efficiently, reducing waste and lowering carbon footprints. Cities will soon benefit as AI optimizes renewable energy sources like wind and solar, making them more viable and cost-effective. AI's contributions to sustainability are helping create a greener, more sustainable planet.

Education, too, is being transformed by AI. Personalized learning platforms tailor educational content to individual students, adapting to their learning styles and paces. Whether in a remote village or an urban center, AI-driven platforms provide access to high-quality education, breaking down barriers and creating opportunities for millions of learners worldwide.

AI in Daily Life

But let's bring it closer to home. Your dishwasher breaks down, and instead of spending hours reading manuals or waiting for a specialist, you use an AI diagnostic app. You record the noise it's making, and within minutes, you get a detailed analysis and step-by-step repair instructions. Suddenly, a daunting DIY project becomes a manageable task. AI isn't just about making life easier—it's about empowering you to tackle challenges head-on and save money.

In the kitchen, AI makes meal planning a breeze. It can generate recipes tailored to your needs, dietary preferences, and even the time you have to cook. It's like having a personal chef on standby, helping you create delicious and nutritious meals while minimizing food waste.

Healthcare, arguably one of the most critical areas, benefits immensely from AI. When you receive blood test results, you no longer need to wait anxiously for a doctor's appointment to understand them. AI can interpret these results, explain their meaning, and suggest questions for your doctor. This empowers you to take an active role in your healthcare, making you an informed participant rather than a passive patient. Such peace of mind is invaluable and can save you significant healthcare costs.

Financial management has become simpler and more efficient with AI tools. Budgeting apps track your spending, categorize expenses, and provide personalized saving tips. It's like having a financial advisor at your fingertips, ensuring you stay on top of your finances effortlessly. Investment platforms analyze market trends and offer tailored advice, helping you grow your wealth with minimal effort. These tools provide financial peace of mind, allowing you to focus on what truly matters.

Personal Growth and Development

AI's role in personal growth and development is equally impressive. Think about learning a new language or picking up a musical instrument. AI-powered learning platforms offer personalized education, providing feedback and support tailored to your progress. This means you can learn at your own pace, in your preferred style, making education more accessible and enjoyable.

Creativity also gets a boost. Writers can use AI to generate ideas, edit drafts, and even create entire stories. Musicians can collaborate with AI to compose music, explore new genres, and refine their work. Visual artists can leverage AI to create digital art, experiment with styles, and streamline creative processes. AI doesn't just automate tasks; it amplifies human creativity, pushing boundaries and opening new possibilities.

When it comes to health, AI offers personalized wellness programs that cater to your unique needs. Fitness apps create customized workout plans, track progress, and provide motivation, helping you stay fit and healthy. Meditation and mindfulness apps offer guided sessions tailored to your needs, promoting mental well-being and reducing stress. By integrating AI into wellness routines, you achieve a balanced and healthy lifestyle effortlessly.

Strengthening Relationships

AI even plays a significant role in enhancing relationships. Language translation apps enable seamless communication with friends and family who speak different languages, breaking down barriers and fostering closer connections. Now, you can plan a family vacation with AI-driven tools that suggest the best options based on your preferences and budget. These tools help create memorable experiences and strengthen social bonds.

For busy families, AI simplifies management by coordinating schedules, organizing activities, and providing reminders. Smart home devices automate household tasks, reducing the burden of daily chores and allowing families to spend more quality time together. AI-powered educational tools support children's learning, making it easier for parents to assist with homework and monitor academic progress.

AI also enhances social interactions. Personalized recommendations for activities, events, and gatherings help you plan enjoyable and meaningful social interactions. Whether it's a weekend getaway, a casual outing with friends, or a family reunion, AI-driven planning tools ensure you make the most of your time together, creating lasting memories.

Familiarity with AI and Responsible Use

Regular use of AI tools provides practical benefits and helps you become familiar with its workings, advantages, and limitations. This understanding is crucial as we navigate AI's ethical and practical implications, ensuring its development aligns with our values and serves the greater good.

By using AI tools regularly, you gain insights into their potential and challenges. This enables you to contribute constructively to discussions of AI ethics, data privacy, and responsible use. This knowledge empowers you to advocate for policies and practices that ensure AI benefits everyone and mitigates potential risks.

Unbelievable Savings in Time and Money

One of the most staggering realizations of integrating AI into your daily life is the unbelievable savings in both time and money. Hours spent on routine tasks are reduced to mere minutes. AI-powered tools can save thousands of hours annually by automating tedious processes and providing instant solutions. For example, using AI for meal planning, home repairs, and financial management can save you over 500 hours yearly. That's over one whole month of waking hours! This freed-up time can be spent on activities that matter most to you, enhancing your quality of life.

The financial savings are equally impressive. AI can reduce utility bills by hundreds of dollars annually by optimizing energy consumption. Smart home devices that learn your patterns and adjust settings automatically ensure you're not wasting resources. AI-powered diagnostic tools for home repairs can save you from costly professional services, potentially keeping thousands of dollars in your pocket each year. Even in healthcare, understanding and managing your medical conditions efficiently with AI's help can significantly reduce medical expenses.

The savings become life-changing when considering the impact of just two or three AI-driven solutions. Whether it's cutting down on unnecessary spending, avoiding costly repairs, or saving on utility bills, the financial relief provided by AI is substantial. These savings translate into more economic freedom, allowing you to invest in experiences and opportunities that enrich your life.

The Future of AI

Looking ahead, AI's potential to transform our lives is limitless. Imagine a healthcare system where AI assists in diagnosis and predictive medicine, identifying potential health issues before they become serious. Picture smart cities where AI optimizes transportation, energy management, and public services, enhancing urban living.

AI will continue revolutionizing learning, making it more interactive, engaging, and effective. Virtual and augmented reality technologies will create immersive learning experiences, making education more enjoyable and accessible to all.

The workplace will also see significant changes. AI will enhance productivity by automating routine tasks, allowing us to focus on more creative and strategic work. Decision-making will be supported by AI-driven tools that provide insights and recommendations based on data analysis, fostering innovation and collaboration.

The AI Revolution

Embracing AI in your daily life is not just about leveraging technology for convenience; it's about empowering yourself to live a more informed, efficient, and enriched life. AI is a powerful tool that, when used responsibly, can enhance every aspect of your life, from the mundane to the extraordinary. As you continue exploring and integrating AI into your routines, you'll find its transformative potential is boundless, offering a brighter, more promising future for you and the world around you. So, step confidently into this AI-enhanced world and let it revolutionize your life, one wise decision at a time.

ABOUT THE AUTHOR

For over twenty-five years, Paul Joffe has consistently contributed to numerous areas of technology, business, entertainment, and education. With a background in philosophy, psychology, and filmmaking, he has spent much of his career developing products that merge technology with an authentic human-centered experience.

In leadership roles at renowned global media companies Disney, Sony Pictures Entertainment, and the BBC, as well as longtime interactive education leader JumpStart, he has led product initiatives that have reached tens of millions of people worldwide.

While he's helped drive many financial success stories, he is most fulfilled by bringing together people from multiple disciplines, diverse backgrounds, and original approaches to work collaboratively and solve problems.

As a professional, Paul is driven by an entrepreneurial spirit. He explores new business models, platforms, and technologies that have the potential to transform industries and foster scale. In his role as COO of Pixelynx, he aims to help the start-up empower the future of artists by merging AI, Intellectual Property, and Blockchain so they can maintain rights over their unique works while receiving fairer commercial returns.

As a creator, he has been honored with multiple Emmy nominations and holds several patents, reflecting a commitment to excellence and invention. He lectured for five years at Loyola Marymount's School of Film and Television, where he also received an M.F.A.

In addition to his professional endeavors, he is a proud member of Abundance360, a community of visionaries dedicated to creating meaningful global impact. As an advisor to Novobeing and Wicked Saints Studios and the founder of Kid Awesome, he is passionate about leveraging technology and ingenuity to drive advancements in education, wellness, and positive social interaction.

Originally a native of Boston, he has two grown children and currently lives in Los Angeles with his wife and two cats.

CONNECT WITH LISA

Follow Lisa Bernard on Instagram for entertaining explanations of AI tools and trends.

@LisaBernard.DemystifyAI

www.ingramcontent.com/pod-product-compliance
Lightning Source LLC
Chambersburg PA
CBHW071545210326
41597CB00019B/3121
* 9 7 8 1 6 3 6 8 0 3 7 7 7 *